BREAKTHROUGH

John Craven and Molly Cox

BRITISH BROADCASTING CORPORATION

The BBC TV series *Breakthrough*
was first shown in March 1981. It
was produced by Molly Cox and directed
by Barbara Kindred

Published by the
British Broadcasting Corporation
35 Marylebone High Street
London W1M 4AA

ISBN 0 563 17936 8

Printed in England
by Jolly & Barber Ltd,
Rugby, Warwickshire
This book is set in Monotype Ehrhardt,
12pt leaded 2pts

Acknowledgements

Acknowledgement is due to the following for permission to reproduce illustrations: BARNABY'S PICTURE LIBRARY modern photographs of: aqueduct, page 18, canal (both Mustograph), page 23, suspension bridge (Walter Young), page 25, Britannia bridge (Don Williams), page 51, Clifton bridge (West Air Photography), page 62. BBC HULTON PICTURE LIBRARY Telford, page 6, Paine, page 16, muddy pavements, page 24, accident, page 49, Brunel, page 52, *Great Western*, page 72, *Great Eastern* saloon, page 81, Suez panorama, page 87, Napoleon & Eugénie, page 90, loading camels, page 95, Ishmaelia, page 97, dredgers, page 98, procession of ships, fête, page 101, ship in canal, page 103, statue, page 104. BOROUGH OF DARLINGTON MUSEUM opening ceremony, page 41. CAMERA PRESS opening ceremony (ILN), page 99. GREAT WESTERN RAILWAY MUSEUM, SWINDON S. Devon railway, page 71. GUILDHALL LIBRARY cross-section of tunnel, page 56. INSTITUTION OF CIVIL ENGINEERS (from "The Works of Isambard Kingdom Brunel", ed. Sir Alfred Pugsley) ship, front cover, tunnelling, page 65; (from "L'Isthme et le Canal de Suez" by Charles Roux) canal cross-section, front cover (redrawn by Corinne & Ray Burrows). IRONBRIDGE GORGE MUSEUM TRUST (ELTON COLLECTION) Eskdale, page 9, Stephenson, page 28, "Sundial Cottage", page 31, Santa Ana, page 37, *Rocket*, page 45, Britannia bridge, pages 50 & 51 & back cover, "Brunel shield", page 57, banquet, page 59, Clifton bridge design, page 63, Paddington, page 65, Sonning cutting, page 67, Swindon, page 69, Saltash bridge, page 80; (TELFORD COLLECTION) mason's mark, page 12, prison, page 13, lock section, page 16, aqueduct, page 17, aqueduct diagram, page 18, Tewkesbury bridge, page 19, & front cover (redrawn by Corinne & Ray Burrows), church, page 20, harbour, page 21, Llynon bridge, page 24, Menai bridge drawings, pages 25 & 26, toll house, page 26, Conway bridge, page 27. MANSELL COLLECTION factory, page 36, *Great Britain* berth, page 75. NATIONAL RAILWAY MUSEUM, YORK (Crown Copyright) poster, page 48. ROYAL COMMISSION ON ANCIENT MONUMENTS, SCOTLAND flight of locks, page 22. SCIENCE MUSEUM, LONDON iron bridge, page 15, colliery locomotive, page 32, seal, page 40, *Locomotion*, page 41, *Rocket*, page 46 (last three Crown Copyright), Tring cutting, Kilsby tunnel, page 47, *Eagle*, Box tunnel, page 68, *Great Britain* aground, page 76, *Great Eastern*, page 79. PHOTOTHÈQUE JULES TALLANDIER de Lesseps, page 82, Saïd Pasha, page 91. DESMOND TRIPP STUDIOS flooded tunnel, page 59. THE TRUSTEES, THE WALLACE COLLECTION, LONDON Eugénie, page 88. WEIDENFELD & NICOLSON (from "The British in the Middle East" by Sarah Searight) Mahomet Ali Pasha, page 85.

INTRODUCTION

The driver gave a warning blast on his whistle and the high-speed train plunged from the bright daylight into the blackness of Box railway tunnel. The filming lights went on, the camera started to run, and I had just over a minute to tell the remarkable story of how this two-mile-long tunnel was built before we were out into the daylight at the other end.

That journey from London to Bristol wasn't new to me – I'd made it dozens of times before, and never given the tunnel a second thought. Like millions of other travellers, I was busy reading, or talking. But this time it was different, because we were filming the story of Isambard Kingdom Brunel, the man who built that railway line and its famous tunnel in the middle of the last century. I began to appreciate, really for the first time, the giant contribution that Brunel, and people like him, had made all those years ago to the lives that we lead today.

Today's trains are running along the same route that Brunel planned so brilliantly for his steam engines. The long straight tunnel which his workmen hacked out of a hillside with pick and shovel is still in use for the high-speed trains of the 1980s.

Brunel was a pioneer, a man of vision – and there were others like him living at the same time. They called themselves civil engineers – a new idea, since before then engineers had been military men, designing guns and other "engines of war". Civil engineers played no part in war; as they said at the time, their task was "to use their art to direct the power in nature for the convenience and use of man".

In the span of one hundred years, from roughly 1780 to 1880, these inventive men changed the entire look of this country, and the lives of the people everywhere. They built good roads, new bridges, underground tunnels; they sent huge steam-ships across the ocean, and they linked those oceans together with ship canals. But above all, they invented and developed steam locomotives, and gave the world its railways.

It's hard to imagine just what ordinary people's lives were like two hundred years ago. There was no television or radio, of course; and no national daily newspapers, and no Post Office to deliver letters. So people found it hard to know what was going on in the world.

The railways hadn't been built, and men walked miles to their work, and children sometimes walked miles to school. The rich had their horses and carriages, but even the main roads were just dirt tracks, white with dust in dry weather and a slithering mass of mud when it rained. In the centre of London, coaches often got stuck with their wheels almost lost beneath the mud.

In the hills roads were dangerously steep. Cart-wheels got smashed and horses fell, trying to drag their loads over rocky tracks. It was hard to travel, and few people did. In the country in winter, village people were entirely cut off. They had to store all the supplies they would need to survive the winter.

For centuries, life had gone on like this, and it took a revolution in technology to change it all. This book is about four of the men who began it . . . Thomas Telford, Robert Stephenson, Ferdinand de Lesseps, and Isambard Kingdom Brunel.

They lived at the same time, they knew each other, they shared each other's problems, they often argued with each other, and sometimes they even fought each other. They were all true pioneers, using their technical skill to make life better for their fellow men.

They took gambles, they had failures, but they never lost their belief in their projects or in themselves. They made big breakthroughs that changed the way the world went about its business; they brought people closer together. And today, the huge changes that they made, almost on their own, still affect all of us in our everyday lives. They were great men.

6 *Breakthrough*

THOMAS TELFORD

I

The Shepherd Boy

The portrait is huge – befitting a man who did huge things. It stands in a place of honour and it shows a proud Victorian gentleman sitting in a black leather chair, one hand resting on an elegant table piled high with papers. In the picture, you can see behind him through an open door, a wooded valley with what seems to be a Roman aqueduct sweeping across it. In fact, it is not Roman – it was designed in 1795 by the man in the portrait, Thomas Telford.

Telford was known as the father of civil engineering in this country. He was the first and possibly the greatest of a new breed of men who changed the face of Britain by criss-crossing it with roads, canals, railways, bridges, viaducts and aqueducts.

He became the first president of the Institution of Civil Engineers, and his portrait hangs today in their headquarters near Parliament Square in London.

The aqueduct in the portrait, with its classic arches, takes the waters of a canal right across the Dee Valley, forty metres above the ground. It was built by Telford near the Welsh village of Fron Cysyllte and today canal boats still travel across it, and people still talk about it – as they did when it was new – as one of the wonders of Wales.

When Queen Victoria was just a young girl playing in Kensington Gardens, Telford was already an old man. He was

Opposite: Thomas Telford's portrait in the Institution of Civil Engineers, painted by Samuel Lane in 1822

honoured and respected throughout the country and famous across half of Europe, not only for the vast canal works he undertook, but for his bold new designs for bridges, and above all for the hundreds of miles of safe, solid roads he had built. Like the Romans his designs were good, and like the Romans he built to last. He was seventy-seven when he died and he was buried with full honours in Westminster Abbey.

But all his life Telford had lived almost like a gypsy, forever travelling wherever his work took him. Sometimes he lived in lodgings, sometimes in local inns; for years his home in London was a couple of rooms above an old coffee shop in the Strand. He was well into his sixties and a rich man before he settled down and bought a proper house for himself. It was an elegant town house near Westminster Bridge and he was proud of its marble mantelpieces and a fine painting by Canaletto that decorated the drawing-room. There, he gave dinners for some of his famous friends, the leading architects, poets, and politicians of the time, and introduced them to the struggling young engineering students that he'd taken under his wing. In fact, he used half of his London house as lodgings for his students. After dinner, over the wine, Telford, the grand old man of Civil Engineering, would entertain the company to stories of his childhood, when he was a penniless shepherd's boy. He was brought up in a village on the banks of the River Esk – a remote corner of Scotland close to the border with England. He remembered those early days in Eskdale with pride and delight.

He was born in 1757. His father was a shepherd who worked at Glendining farm up in the hills, but Tom Telford never knew him. He died soon after Tom was born. So Tom's mother moved back down the valley to live in one room in a small cottage near the church at Westerkirk. One of her brothers taught Tom to work with sheep like his father and in exchange provided him with boots. Another uncle, a farm-worker at the Johnstone's big house by the river, urged his mother to let Tom go to school. In those days, there was no law that said children had to go to school

– many poor children didn't get any proper education, they went out to work as soon as they were old enough and strong enough. But in Scotland each village managed to provide its own small school. It cost money to go, a penny a session for lessons, and that money had to be earned.

So Tom ran errands for his uncles, helped them at lambing time and spent hours scaring the birds away from the fields. He enjoyed the village school. It was an old-fashioned place where farm-children, the children from the big house, the doctor's and teacher's children all learnt to count and read and write together in one big room. Tom was known as "laughing Tom", and the friends he made at school were to remain his friends for life: the Johnstones, the Littles, the Davidsons, the Elliots and the Jacksons. Like the Telfords they were all old Border families, independent people whose ancestors had once banded together on raids across into England to steal cattle. Many years later, it was their grandsons who became Telford's engineering pupils and shared his London house – and sat dutifully listening to stories about his early life in Scotland.

In Thomas Telford's young days, the villagers of Eskdale had a great respect for books and learning. Long before there were any free public libraries, Westerkirk had collected a small library of its own. It opened one night a month, at full-moon, to give the men who lived away up in the hills a chance to walk down after work, collect a book and get back home safely by the light of the moon. The book had to last them for four weeks till the moon was full again.

In Eskdale

At school Tom loved reading and most of all he loved reading poetry. What he really wanted to be was a poet like his hero, Robert Burns. While he sat on the bare hillside guarding sheep he would try scribbling down his own poems. He wrote this for one of his cousins:

"We, my friend, will often steal away
 To this lone seat, and quiet pass the day;

> Here oft recall the pleasing scenes we knew
> In early youth, when every scene was new."

But poetry wouldn't keep him in boots; he had to learn a proper trade. There was no work to be had in the valley, no chance to be a shepherd like his father, or a farm-worker like his uncle, so at the age of fourteen Tom Telford set off to walk to the nearest town. His childhood was over.

He became an apprentice to a stone-mason, but his master was so cruel and beat him so often that Tom was forced to run away home. But he couldn't stay with his mother for long, so he decided to go to Langholm, the local market town, where there was a lot of new building work, and this time he was determined, given half a chance, to learn to be a first-class stone-mason.

He was luckier with his new master. The apprenticeship was four years' hard work, and after he had helped to complete the bridge over the river at Langholm, he was allowed to put his own stone-mason's mark on one of its piers. This meant his training was over at last. The bridge stands to this day.

Though he was now a fully-trained craftsman, he still had time for books and poetry. He made friends with an old lady in Langholm, a Miss Pasley, who allowed him to borrow any book he wanted from her library shelves. So sometimes young Telford the stone-mason could be found carrying a carefully wrapped-up book of poems among his bundle of tools when he set off to work to repair houses in Langholm. Sometimes he went to Westerkirk, where he worked on the local manse. "I read and read and glowered; then read and read again", he said later, remembering how he had struggled to understand Milton's long poem *Paradise Lost*.

A few years later, he decided to seek his fortune in England, but before he left Scotland, he placed a plain square-cut gravestone in Westerkirk churchyard. It was to be a memorial to the family he had never known, for a baby brother also called Thomas who had died before he was born, and for John Telford his father. You can still find the stone and read the carefully chosen words: "In memory of John Telford who after living thirty-three years as an unblameable shepherd died at Glendining 13th November 1757; and his son Thomas who died an infant."

Then Telford, at the age of twenty-five, set off to travel the long hard road to London, on a horse lent to him by the Johnstones of Westerhall and wearing breeches borrowed from a Jackson cousin. His most precious possessions, his stone-mason's tools, were wrapped in a workman's leather apron, together with some letters of introduction from the good Miss Pasley to her London friends. He was sad to leave Scotland, and when he reached London he wrote home to Andrew Little:

> "How is Andrew and Sandy and Aleck and Davie? Remember me to all the folk at home. Tell that glutton-boy John Elliot to behave, or come down here and I'll fight him. Let Matthew Davidson know that I enquired at Carlisle after the spy-glass he wanted, but the man had none. My compliments to that lassie Jenny Smith. Let my mother know I am well and that I will print a letter for her soon."

His mother could scarcely read so Telford always printed her letters to make it easy for her. He wrote regularly to his best friend Andrew Little, but tragically, Andrew could not see to read them – his brother had to read everything for him. After school Andrew Little had trained to be a ship's doctor, but on his very first voyage, in a tropical storm, he had been struck by lightning and was blinded. Telford wrote to him for years, long letters like a diary telling him all about his new life in the south; the people he met, the work he did, and his ambitious hopes for the future.

> To Andrew Little, *London* July 1783: "At present I am laying schemes of a pretty extensive kind. My vanity is apt to say when looking on the common drudges, here as well as in other places, 'I was born to command ten thousand slaves like you!' This is too much; but at the same time it is too true, for I find the workmen here to be more ignorant than they are in Eskdale – there's not a

Stonemason's mark on Langholm bridge – Telford's first recorded work

Matthew Davidson among them! Mr Adams is uncommonly kind. Sir William Chambers needs to be consulted too. Somerset House might be the finest opportunity – it would flatter my vanity much, to have it said that I had a hand in that noble work. There is nothing, but just having a Name here to make a fortune."

Sir William Chambers did let him have a hand in building Somerset House, but it didn't make a name for him, or a fortune; he was engaged as an ordinary stone-mason. At that time London was in a flurry of new building schemes; the Adams brothers were designing their fine classical town houses; the crescents, squares and wide streets of Regency London were being planned. It seemed to young Telford that to make a name he needed to learn to be an architect. So he spent his evenings studying the classical books on architecture. He was prepared to go without many of the ordinary comforts of life, but he was never without books.

By the next year Telford was in Portsmouth, no longer a stone-mason but Superintendent of the new buildings in the Dockyard. He was already boasting that he had his hair fashionably powdered every day and wore a clean shirt three times a week! But he continued to fill his evenings with studying, as he told his friend Andrew:

> "You ask me what I do all winter. My business requires a great deal of writing and drawing. Then, knowledge is my most ardent pursuit, and I am not contented unless I can reason on every particular. I am now very deep into Chemistry; having looked into some books I perceive the field is boundless. I am determined to study until I attain some general knowledge of Chemistry as it is of universal use in the Arts as well as in Medicine. Often my dear Andrew do I wish to have you along with me. You would find me surrounded by Books, Drawings, Compasses, Pencils, and Pens; great is the confusion but it pleases my taste and that's good enough."

By the time he was thirty Telford had become the enthusiastic and energetic County Surveyor of Shropshire, and was in charge of all the roads, bridges, churches, hospitals and public buildings of the county. There's a story he liked to tell of how he became known all over Shrewsbury and throughout Shropshire. There was an old church in the centre of Shrewsbury town, St Chad's. For years the church roof had been leaking badly but nobody wanted to spend money on major repairs. They asked Telford to take a look at it and give his opinion. He examined the place and swiftly reported back to the Churchwardens. It wasn't just the roof that needed repairs, there were dangerous cracks in all the church walls. At any moment, he warned, St Chad's might collapse. The Churchwardens didn't want to hear such bad news so they persuaded some local workmen to do a patch-up job on just the roof. On the morning the men arrived, they waited with their ladders and ropes for the Sexton to bring the key and open the church door. Suddenly, the clock in the tower struck, and all the church walls came tumbling down! Within minutes old St Chad's was a magnificent heap of rubble. The good people of Shrewsbury were amazed; Telford's words had come true.

One of the public buildings Telford was strangely proud of was the new prison in Shrewsbury. And by chance, while it was being built he met John Howard, the great prison reformer, and was

Shrewsbury prison as it is today. The bust over the doorway is of John Howard

completely won over by the extraordinary personality of this little old man whose only wish in life was to stay quietly at home, yet who spent all his days travelling the world preaching in every country the need for a more humane way of treating prisoners. Telford was so impressed by this "Guardian Angel of the miserable and distressed" that he persuaded the prison governors to allow gangs of prisoners to go out to work on some of his new projects. He was at the time excavating the remains of a Roman city at Wroxeter, a very early piece of archaeological work. With the help of the prison-gangs he uncovered Roman baths and tiled floors and many pillared buildings. He made careful drawings of them all and wrote descriptions "so that men of learning might satisfy their curiosity".

Telford kept amazingly busy. He visited London to read about antiquities in the British Museum; he visited Oxford and admired its colleges and churches; he went to Birmingham to buy stained glass for the church in Bridgnorth he was building. But he didn't think much of Birmingham, he called it "that place famous for Buttons and Buckles, Ignorance and Barbarism". He made many visits to Coalbrookdale in Shropshire, where the first bridge in the world made entirely of iron had just been built. And he made a point of getting to know the iron-masters and visiting their foundries to study at first hand the new "cast" iron. It was a material he was to use himself over and over again.

Although he was beginning to be prosperous, he still lived a simple life in lodgings.

"These last six months I have taken to drinking water only. I avoid sweets, and never eat any nick-nacks" he wrote to his friend Andrew. At the same time he sent him a sum of money to be shared between him and Telford's mother.

"I have the opportunity of earning money. You have not and therefore I insist on sharing a little. I don't wish anybody to know of this besides yourself. To tell you the truth, to set my mother and you above the fear of want has always been my first objective, although I have never told you so before. And next to that, to be

The first cast iron bridge in
the world, completed in
1779. It still spans the River
Severn at Coalbrookdale

that somebody that you have always taught me to believe I have a
right to. And I humbly presume there is something in it – it may
be self-confidence, but I think wherever I was, there has always
been a bustle, a scene of contention, and I have always been the
prominent feature."

He was about to be the centre of another more serious "scene of
contention" – one in which his bright career nearly came to a
sudden end. His interest in the use of iron for making bridges
took him to London, where the decorated iron work for a huge
new bridge, one hundred and twenty metres long, was on display
on Paddington Green. It had been made in Yorkshire, to the
designs of an American, Tom Paine. It was to be used to bridge
the Schuyhill River in Pennsylvania. But before Tom Paine and
his fantastic iron bridge could take ship for America, he was
arrested and sent to prison. The authorities didn't like Paine or
his ideas. He had written a book, a very revolutionary book called

The Shepherd Boy 15

Contemporary engraving of
Thomas Paine

The Rights of Man, and although the American revolt against the British was over, a much bloodier revolution in France had just begun.

But Telford was so interested in the bridge and in Tom Paine's ideas, he sent a copy of *The Rights of Man* to the blind Andrew Little in Langholm. Of course, it had to be read out loud to him and it caused quite a stir among the young people of Langholm. Telford was immediately accused of starting a revolution in Scotland! As a County Surveyor, he was a public official and for a while his future employment was in danger. On his next visit home to see his mother, he told his friends he always kept his working clothes at the ready, and his stone-mason's chisels sharp, in case he needed them. "You never know," he said, "what might happen next."

But what did happen next was quite unexpected. Instead of losing his job, he was offered another and much more important one, at a salary of £500 a year – a lot of money in those days. He was to be Principal Assistant, Surveyor and Architect to William Jessop, to work on a great new canal network – the Ellesmere Canal. All over England industrialists and manufacturers were building canals to carry their goods to and from the cities and ports. Barges full of coal, timber and iron were towed by horses from one end of the country to the other. To build a canal across flat country is fairly simple, but to take one across hills and wide valleys means building many locks, tunnels and bridges. The new Ellesmere Canal project was to link the iron foundries of Shropshire with the coast. From the steep valley of the Severn, its course would run across the Welsh hills to the flat lands of Merseyside, and the sea. It was a great challenge, and Telford was excited by the prospect. He wrote home:

Cross-section of a lock on
the Ellesmere Canal

"I believe it is the greatest work in hand in the Kingdom; it will require great exertions but it is worthy of them all. There are to be bridges over several rivers which cross the line of the canal, several locks, two aqueducts and a tunnel underground. I have just recommended iron for the very great aqueduct over the River

Nineteen arches support the Pont Cysyllte aqueduct over the River Dee

Dee. It will be executed under my direction upon a principle entirely new."

The "very great aqueduct" over the River Dee known in Welsh as Pont Cysyllte was to make Telford famous. It was in all, a truly enormous project, and to help him with his work Telford sent for his old friend from his Eskdale days, Matthew Davidson, to be his Superintendent. When William Jessop saw the height of the aqueduct marked on Telford's plans, he said he was afraid workmen would turn giddy with terror working forty metres up in the air. But they didn't – in fact, in the ten years it took to put the iron-bound canal on its stone piers across the Dee valley, there was only one accident.

Above: "The Navigable Aqueduct of Pont-Y-Cyssylte for the Ellesmere Canal over the River Dee at the Bottom of the Vale of Llangollen." One of several hundred drawings recorded in the *Atlas to the Life of Thomas Telford, Civil Engineer*, published in 1838; *right*: the aqueduct today

Telford was a man who cared for every small detail of his work. The giant earthworks which form the long embankment on the south side of the valley were planted with carefully chosen trees, to hold the soil in place. The tow-path, the delicate iron railings, the specially constructed stone-piers, the bolts for the cast-iron plates that kept the canal water-tight; everything was carefully thought out, designed, measured and tested. Sadly for Telford, before the work was completed, his old friend Andrew Little died. He kept in touch with Andrew's family, but it wasn't the same.

When the aqueduct was finished, on the grand opening day, a procession of six canal boats set out across what was known then as "the stream in the sky". In one of the boats a military band played *Rule Britannia*, and in another sat Thomas Telford, Matthew Davidson and his wife and family. Below in the valley thousands of people stood and stared up in wonder, while a salute of fifteen guns echoed round and round the Welsh hills.

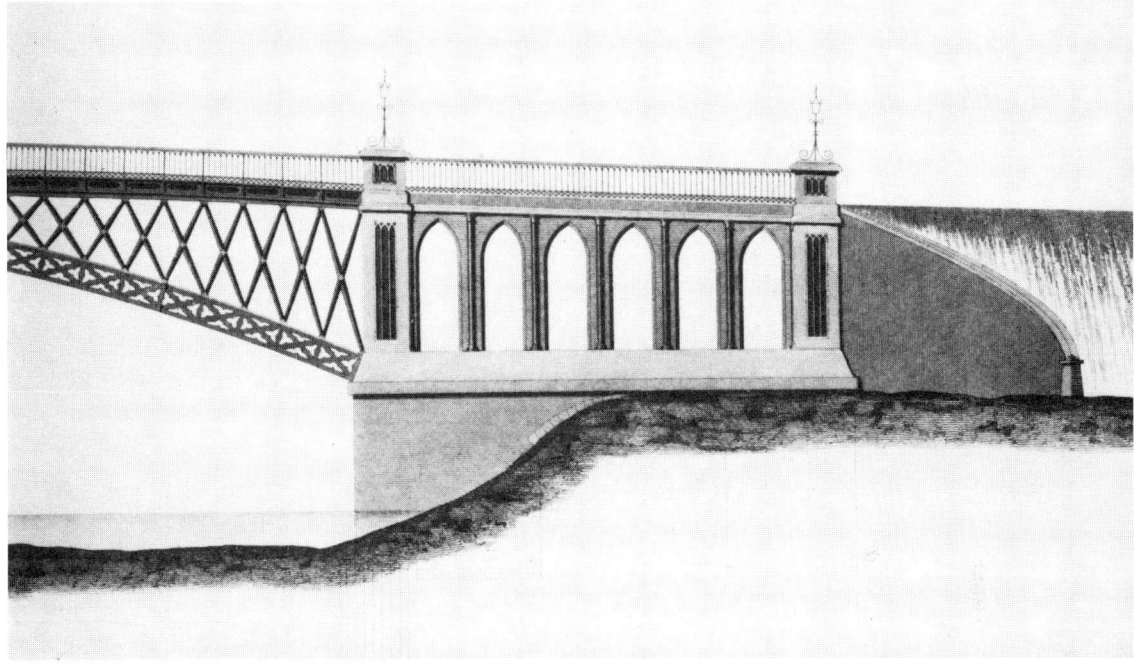

Every engineer has his favourite piece of work. I think Pont Cysyllte must have been Telford's. After it was completed, he continued to visit it regularly; indeed, it's said, he came to take a look at it every year till he died.

But Telford was already deeply involved in at least a half-dozen other major projects. The River Severn had burst its banks and there had been a serious flood. Bridges had been swept away, and new ones had to be built at Buildwas, Bewdley, Bridgnorth, Tewkesbury and Gloucester. And in London he'd been working

Detail of Tewkesbury bridge over the River Severn in Gloucestershire

Church of St Mary at
Bridgnorth, Shropshire,
designed by Telford

hard on detailed plans for a new iron bridge, a single span to cross
the Thames, to replace old London Bridge with its tumbledown
arches. Alas, his plans were never used.

He worked on a new form of underwater cement, the result of
his studies in chemistry. He was commissioned to make a report
on industry and communications in the Scottish Highlands.

During two long summers he toured these desolate areas,
where until recently the old Highland clan system had operated
well. But since the Jacobite wars fifty years before, the people
were without hope and without work; the farmlands had been
given over to sheep, the villagers were emigrating in their
thousands to America. Something needed to be done. Telford's
report suggested that the most pressing need was for bridges and
main roads to connect the north of Scotland with the industrial
cities in the south. Then new safe harbours for the fishing fleets
were needed, at Banff, Peterhead and Aberdeen. But his boldest
plan was to put forward an old idea, one that many engineers had
studied but none had had the courage to start. It was to build a
canal which would join together the lochs of the Great Glen –
including Loch Ness – and make a continuous waterway right
across the centre of the Highlands, to join the North Sea in the
east to the Atlantic in the west. It wasn't to be a canal for barges,
but sea-going ships – Britain's first ship canal, the Caledonian
Canal. Naval vessels, trading ships and fishing boats would have a
safe passage inland across Scotland, instead of making the long
and dangerous journey round the wild, windswept coast of
Northern Scotland: that three hundred miles of rocks and mists
and cross-tides which had once wrecked the whole Armada fleet.

All Telford's plans for the Highlands were agreed. They would
provide much-needed work for the Highlanders and teach many
of them new skills.

Even before the Ellesmere Canal scheme was completed,
Telford had already worked out many of the details of the plans
for the Caledonian Canal. Indeed, hardly had Matthew Davidson
finished his triumphant journey across Pont Cysyllte aqueduct

One of Telford's safe new harbours in Scotland, at Dundee

than he and his family were whisked away to Inverness so that he could supervise the building of the east end of the new canal and in particular the building of the entrance, the sea-lock at Clachnaharry.

When work on the Caledonian Canal started, we were still at war with Napoleon, the Navy still used Nelson's great wooden sailing ships, steam-engines had only just been put to use and most work had to be done by hand. More than three thousand men worked on building the canal, every year for nearly twenty years. Some of those men lie buried in little cemeteries up in the hills near where they worked and died. New roads, bridges, and locks had to be built. Each lock-gate was specially made of iron, to withstand sea water, and weighed fifty tons. These gates had to be carefully swung into place. At Banavie, at the west end of the

The Shepherd Boy 21

canal, a flight of eight locks drops the canal water down twenty-seven metres to sea level in just over three kilometres. It's a spectacular sight. Telford himself christenend it "Neptune's Staircase". Matthew Davidson worked on the canal till he died, then his son James took over, and in his turn his son John became supervisor. It's a tradition, even now, for whole families to continue to hand down the working of its twenty-eight locks and eight swing bridges.

Making a journey through the Caledonian Canal today is a

"Neptune's Staircase" – a flight of locks on the Caledonian Canal at Banavie

vivid reminder of the personalities, the effort, and the imagination of the men who built it. When it was opened and the first ship sailed through from sea to sea, it had cost £1,000,000 – far more than Telford had planned. But it's still used today, a hundred and sixty years later, and it's still a safe sea route for fishing boats to cross from one side of Scotland to the other.

All the time it was being built, Telford himself was continually travelling up and down the long line of the canal. At the same time, he was supervising the making of new harbours for Peterhead, Aberdeen and Dundee, and designing bridges for Dunkeld, Cartland Crags in Lanark, Craigallachie, Carlisle, at

Bonar near the Dornoch Firth, at Wick, Aultmore and Cannock and many, many more. He was also building more than nine hundred miles of main roads. All his travelling was done on horseback or in a pony and trap, and a lot of it over wild country in bad weather. He was never able, nor ever wanted, to take a holiday. He planned a new ship canal for the King of Sweden and a road from Moscow to Warsaw for the Tsar of Russia.

In fact his friends complained that he was in love with his work. He took one of them, the poet Robert Southey, with him on one of his long Highland journeys, and Southey kept a diary of all he saw. He wrote:

"The Effects of rough Pavements to tender-footed Passengers" – a cartoon by Cruikshank published in 1797

"Telford's is a happy life; everywhere making roads, building bridges, forming canals and creating harbours – works of sure, solid, permanent usefulness, and everywhere employing a great number of persons. A man more heartily to be liked, more to be admired, I have never met. The plan on which he makes a road is this: to level and drain it, and then like the Romans, to lay a solid pavement of large stones, as close as can be set: then a layer of stones broken to about the size of walnuts; and over all a little gravel, if it is to hand."

Like Roman roads, Telford's roads were built to last. Some of them we still use. The road to Holyhead, which takes the main traffic to Ireland – the A5 – still follows the line Telford first put down all those years ago. It goes through the rocky valleys of North Wales and at pretty Betws-y-Coed there's Telford's

Bridge at Betws-y-Coed – from Telford's *Atlas*

Waterloo bridge across the River Conway. It was built in 1815 and celebrates the defeat of Napoleon at Waterloo in beautiful coloured iron-work flowers. Further along the same road is the old walled town of Conway with its ancient castle. There Telford built another bridge across the estuary, and decorated it to match the old castle walls.

But the most famous of all of Telford's bridges is a few miles further on and joins the island of Anglesey to the mainland of Wales – the Menai suspension bridge. Before the bridge was built, all the traffic for Anglesey was ferried across the Menai Straits in open boats. But it was a dangerous crossing; in this narrow channel of sea, the current runs fast, the tides change

A nineteenth-century artist's impression of the Menai bridge

quickly and storms from the surrounding hills arrive without warning. Telford's bridge was to be the first attempt to make a roadway across to the island, but it had its problems; it had to be built high enough above the water to allow ships to pass beneath it, and it had to be strong enough to withstand the winter gales. Telford decided to build his bridge thirty metres up above the water, to make it a simple dual carriage roadway 183 metres long, suspended from two tall towers by iron chains. It was something that had never been done before. Foot-bridges hung from ropes or iron chains had been built before, but the idea of a road suspension bridge that size was quite new. Everyone said that it wouldn't work, that it was a mere castle – or rather bridge – in the air. But Telford had worked out carefully all he wanted to do. He had sixteen great chains made, each weighing twenty-three and a half tons, and all of them made of the very best iron by his old friends from the iron-foundries of Shropshire. He tested and re-tested the chains; when he was quite sure they were all strong enough, he then had to find some way to put them in place from one side of the water to the other.

Menai Suspension bridge today

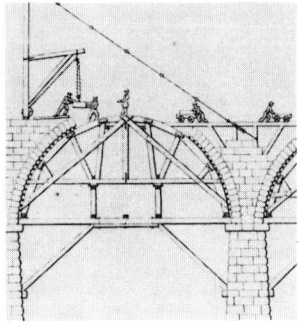

Drawings published in a book by W. A. Provis, who worked on the construction of the Menai bridge. *Top*: one of the piers; *below*: construction of the arches

Toll house on the Holyhead road, from Telford's *Atlas*. Several of these stand beside the modern A5 road

The day came when everything was ready – the first chain was floated out on a special long barge across the water at high tide. From each side a hundred and fifty men hauled on ropes. Then, as the tide changed and the sea water slowly ebbed, the long barge floated away, leaving the great chain swinging free. The first link across the Menai Straits had been made. A great cheer went up. Telford had hardly dared to look – he was on his knees praying. But the worst was over, and his excited workmen clambered like monkeys along the swinging chain – cheering, laughing and singing. Four months later, when all the chains were fixed in position, a Welsh brass band marched solemnly to the centre of the still-uncompleted bridge and played the National Anthem, while the steamboat *St David's* passed to and fro below to the cheers of the local people. It was their own spontaneous celebration. Within months, in January 1826, the Menai Suspension Bridge was more formally opened to be used by public traffic – as it is still used today.

If Telford's great aqueduct was a stream in the sky, at Menai it was a bridge floating in the air. It was a huge success, and people all over the world have been building similar suspension bridges ever since. But of all his bold projects, the Menai Bridge was the only one to give Telford sleepless nights of worry. He was getting old: in fact he was seventy years old. His world was changing around him and he found it difficult to understand. Already a new breed of young engineers were building the new-fangled railways; the age of steam was starting and he felt a little left behind.

When he was seventy-six, the year before he died, he took part in an experiment for the use of self-propelled steam-carriages on the main road between London and Holyhead. He travelled in one of them himself halfway to Birmingham before one of the tubes in its boiler burst, and the carriage could go no further. It's odd to think of that old gentleman all those years ago trying out a self-propelled carriage on the road to Birmingham – like an ancestor of the modern car trying out the M1!

When Telford died he left his books to the Institution of Civil

Engineers for the use of their students, but he left money to set up a new free library in Langholm, and more money to buy books for the small village library of his old home in Westerkirk. The friends of his childhood and books were the things he loved most in his life.

Though he is buried among the famous in Westminster Abbey, it was his friend Robert Southey who wrote his memorial – a poem cut in a marble slab, which stands beside the offices of the Caledonian Canal:

> "... may the marble here
> Record the Architects immortal name.
> TELFORD it was by whose presiding mind
> The whole great work was planned and perfected ..."

A poem for Tom Telford, the shepherd's son, who had created so many admirable and enduring works of engineering, but who had really wanted, more than anything else in life, to be a poet himself.

Robert Stephenson.
Britannia bridge over the
Menai Straits is in the
background

ROBERT STEPHENSON

2

The Rocket Builder

There's not much left to mark the spot where history was made – just a plaque on the wall. The shed is now a storeroom for a builders' merchant, and shafts of light stream down from the rafters on to piles of wheelbarrows and lavatory pans.

But half close your eyes, let your imagination take over, and you can picture what it must have been like here more than a hundred and fifty years ago. All the noise, activity and thrill as the world's most famous railway engine began to take shape. For here in Forth Street, close to the quayside at Newcastle-upon-Tyne, the *Rocket* was born. It was the pride of Britain's greatest railway engineers, Robert Stephenson and his father George, and it was to start a total revolution in the way that people travelled.

A few months ago, I took a ride on *Rocket*, or rather on the beautiful replica that belongs to the National Railway Museum, on a private line at Didcot in Oxfordshire. Standing on the open footplate, the cold air blasting my face and smoke belching from the tall chimney on its barrel-like boiler, it was easy to understand what all the fuss was about when *Rocket* first took to the rails.

To many people, the "iron-horse" was more of an iron monster. "I would rather meet a highwayman or a burglar on my estate," stormed one fox-hunting colonel, "than one of these new-fangled railway engineers." Another noble gentleman exclaimed:

"If this sort of thing is permitted to go on, the nobility will be destroyed!"

But despite these protests from the privileged, the dawn of the railway age could not be held back. For millions of ordinary people, it meant that for the first time they could travel quickly and quite cheaply all around the country, and fifty years after *Rocket* steamed out of the factory in Forth Street, Newcastle, the tentacles of the railway had stretched right across the world.

It was men born and brought up around the British coalfields and iron foundries who invented railway engines and gave England pride of place in the story of locomotive engineering. Men like the Stephensons.

In 1803, when Robert Stephenson was born, England was at war with France. The French Revolution, with its cry of "Liberty, Equality and Fraternity", was scarcely over and people all over Europe were still reeling from its consequences.

But as well as fighting the French, the British were also in the middle of a revolution of their own. It was a very different kind of upheaval, which turned farmland into factories and made Britain into the manufacturing nation of the world – the Industrial Revolution.

All over Wales, in Scotland and in the North of England, fields were spoiled and blackened by coal-mines, iron foundries, furnaces and mills. Beside the rivers small collieries were surrounded by heaps of slag, coal-dust and ashes. The sounds of this countryside were the sounds of pumping-engines and steam-hammers. By day the iron furnaces let out dense smoke and jets of steam. By night lurid flames lit up the sky. Men and boys who had once been farm labourers toiled night and day in the foundries and mines.

When the owners decided a coal-pit had given all it could, the workers were moved, lock, stock and barrel, to the next one and then on to the next, and with them went their families.

George Stephenson was one of those workers. He was only eight years old when he first went to work in the pit at the village of Wylam on the River Tyne. By the time he was twenty-one he had been moved several times and at last found himself at Long Benton, just north of Newcastle-upon-Tyne, in charge of the pumping-engines at Killingworth colliery.

He was married and his baby son, Robert, was just two years old. They lived as all miners did in a cottage owned by the colliery. No. 1 Paradise Row was a stone-built cottage; although there were fields and farmlands around, the road to the pits ran alongside the cottages and it can't have seemed much like Paradise! But George made more money than most, adding to his wages by mending clocks and repairing boots, and saving costs by making furniture for the cottage.

Soon after the family moved there, George's wife and baby daughter died from tuberculosis. George's sister came to look after the house, but George spent much of his time looking after his son Robert, afraid that he might also get sick.

Years later, Robert sometimes complained that he was left-handed because his father carried him round as a baby with only his left hand free!

The little cottage is still there, standing now in the shadow of Killingworth New Town, and it has become a place of pilgrimage for railway enthusiasts from all over the world. But now it's called Sundial Cottage.

The cottage with the sundial over the door

Above the doorway hangs a white sundial with the date August 11 1816 painted on it in black. It was put there by Robert while he was still a schoolboy. He was interested in sundials and wanted his father to make him one. George said he'd do it if Robert worked out the correct angles of the sun.

But Robert thought the mathematics and the astronomy would be too difficult for him to tackle. His father said: "It's to be done, so just get on with it." In fact, that could have been George's motto for living. So it was done – no argument – and the Stephensons' home-made sundial is still there to this day.

Robert was sent to school with other colliery children a mile and a half away at Benton school – a small place where eight boys

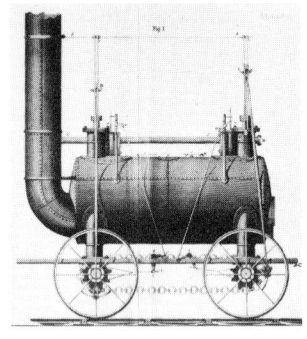

did their lessons on the ground floor, and upstairs eight girls did theirs. It cost his father a penny a week.

Like other miners' children, Robert also had a job. In the mornings, before he went to school, he carried pickaxes to the blacksmith's to be sharpened, and brought them back to the pit on his way home. At most mines, horses were used to pull wagons full of coal along tramways to the nearest stream or canal. There, the coal was loaded into barges and towed – again by horses – to the coal staithes at Newcastle for shipping to Hull, Harwich or London.

But at Killingworth colliery, George Stephenson, now a skilled enginewright, built special engines that travelled along tracks pulling the coal-wagons. He gave them names, just as the horses had names; one was called Blücher, in honour of the Prussian General who fought at Waterloo, and the second was named Wellington, after the British hero. *Wellington*, the steam-engine, went on pulling coal-wagons for another fifty years. People admired these "steam-elephants" as they called them, and soon George was being talked about by the coal bosses at Tyneside.

They offered him better jobs with more money, but George felt embarrassed – he had never been to school and he couldn't read or write or set down sums. He had once tried to teach himself from a book but he didn't get very far, and he was not able even to sign his name.

He was determined that everything was going to be very different for his son. He took Robert away from the village school and sent him to the best Academy for young gentlemen in Newcastle. It was five miles away, too far to walk, so his father bought Robert a donkey to ride to school. Robert was very proud of his donkey but he got teased by the other boys for his rough "Geordie" accent, his hob-nailed miners' boots, and for the simple lunches of bread and cheese his aunt wrapped up for him.

Every evening when he got home Robert had to teach his father all he had learnt from his school-books. Then George would tell him all he knew about mechanics; how the Watts engines that he'd seen used in Scottish mills worked; how the great Cornish mine engineer Richard Trevithick had once visited Wylam and how he had invented a steam-carriage; how his own colliery engines worked and how the tracks they ran on could be improved.

Whenever Robert wanted to go out to fly kites or go birds-nesting with other boys, his father kept him working at his books. People said he was too strict with Robert, but George Stephenson had good reason to want his son to grow up to be able to hold his own among educated men.

George had invented a safety-lamp for miners; it was so good, it was used in all the local pits, and by way of thanks the Newcastle mine-owners gave him £100. At precisely the same time a famous London scientist, Sir Humphry Davy, brought out an almost identical safety-lamp. It was soon the talk of London, and mine-owners all over the country awarded Sir Humphry £2000 in recognition of his work for the safety of miners.

George thought this was unfair and wanted to write a letter to complain, but he couldn't write. So Robert wrote down the words that his father dictated. The end result was that George was presented with £1000 and an inscribed silver tankard at a public dinner at the Newcastle Assembly rooms. To his embarrassment, he had to make a speech of thanks and this is part of what he said:

> "When I consider the manner that I have been brought up and lived, and when I consider the high station of Sir H. Davy, and his influence on scientific men and bodies; it lays me under a Debt of Gratitude to the Gentlemen of this meeting which will remain with me so long as ever I shall live."

After his father died Robert kept that silver tankard on his own sideboard to remind him of those early days. And when he was asked who really invented the first miners' safety-lamp, he would say: "If my father hadn't invented it, Sir Humphry would have done so, and if Sir Humphry Davy hadn't invented it, then my father would have done so."

When Robert was sixteen he left school and George married again. Robert was a witness to the wedding and this time when George signed the register he was able to write his full name properly – not just put his mark.

His wife Elizabeth was a loving stepmother as well as a good wife and altogether his father's marriage was a great relief for Robert; the long evenings spent poring over books – for his father's sake – could stop at last.

By now George Stephenson's colliery engines were beginning to be known far beyond Newcastle. There were powerful fixed engines which could haul a string of coal-wagons uphill by ropes, and there were the lighter travelling-engines, "locomotives", like the *Wellington*, which pulled them along rail-roads on flatter ground. Engineers from France, Germany and Russia came to Killingworth to look at them and went home to write books about them. The railway revolution was about to hit the world, and the Stephensons were at the heart of it. It all happened very fast, almost too fast, and inevitably sometimes mistakes were made.

In 1821, Edward Pease, a rich Quaker who lived in Darlington, put a startling proposal to George Stephenson. Instead of building themselves a new canal, like everyone else had been doing, the men of Darlington wanted a railroad, to send the coal from their mines across to Stockton and the sea. Could George build them one?

It was to be like a private colliery line, but twenty-four miles long and all the local coal-pits would pay to use it. George thought it was a great idea – he couldn't wait to start work on what would be the first railroad for public use. Now that the seventeen-year-old Robert was trained and able to draw up plans, they could make a proper survey. They could go further than that; George could get the right sort of rails made for the track in a local iron-foundry. If Mr Pease and his friends wanted, all the coal-wagons on the new railroad could be hauled the whole twenty-four miles to Stockton by steam-engines.

They could use fixed engines for the hills and locomotives for the flat ground; indeed given the money and a factory to make them in, the Stephensons between them could design and manufacture whatever engines were needed. All Edward Pease and his colleagues needed to do was put up the money. Pease was impressed and wrote to one of the Directors in his quaint Quaker style:

"The more we see of Stephenson, the more we are pleased with him. He is altogether a self-taught genius. Don't be surprized if I should tell thee, there seems to us after careful examination no difficulty in the future of laying a railroad as far as London to Edinburgh on which waggons would travel at the rate of twenty miles an hour."

Edward Pease seemed to have caught railway fever long before a single rail had been laid for his new railroad, or even a suitable locomotive designed.

George was pleased to have Robert working with him on the survey. He wanted the calculations to be faultless, and in fact, when they had completed their work, the plans for the proposed railroad were signed *Robert Stephenson, Engineer*. The next step was for the survey to be approved by parliament. Robert went off to University in Edinburgh, but he stayed there for only one term – his father needed him. At Darlington, to work out some calculations. In London, to explain the railroad survey to some MPs. In Newcastle, to start the factory in Forth Street for building steam-engines and coal-wagons. When people asked George why he didn't give Robert a proper university education, his reply was: "I don't want my son to be a Gentleman – Robert must work, work, work; as I have done before him."

George by now had another and even more important railway survey to carry out at Liverpool, and he needed Robert there. Robert was urgently needed at Newcastle to supervise all the new buildings for the factory; to find the skilled staff, order equipment, draw up plans, and make estimates. It was the first factory ever built for the manufacture of steam-locomotives.

Forth Street factory, Newcastle-on-Tyne – the first workshop designed for building locomotives in the world

Robert was managing director at Forth Street – and he was not yet twenty-one years old. It was an enormous responsibility for such a young man, and made more difficult by his father's constant demands. All the new engine designs were meant to be supplied by George, but George was often too busy to do them. He had got involved in yet another new railway project, this time in the Midlands. He sent for Robert again.

"I am very much in want of Robert," he wrote, "you will send him off as soon as possible as I want him to go to Knaresborough and also to do business on the Darlington Railway . . ."

Quite naturally, Robert was beginning to get rather tired of always being at his father's beck and call. He wanted time to live his own life. He had a girl-friend in London, the daughter of a city merchant; he wanted time to see her. He wanted to run the Forth Street factory his own way with the chance to try out his own more professional ideas about engineering.

To show his independence, Robert travelled to Cornwall to witness for himself the different ways in which engines were used in the mines there. He wrote to his father explaining how good it was for him to see new sights and examine new ideas, and how great were the benefits of getting away from home for a while.

Robert Stephenson's cottage at Santa Ana

This letter should have been a warning signal to George, because for months, Robert had been making secret plans to get *right* away from home. He had been offered an exciting new job in Columbia in South America – his mission was to re-open some ancient silver mines. In those days South America was an unknown land filled with riches just waiting to be discovered. Stories were told of its vast rivers and huge rain-forests, its rare animals and exotic plants. At the silver mines at Santa Ana, Robert would be his own master. He could choose his miners and order for himself the machinery and stores he wanted. All the decisions would be his alone, and the prize at the end could be a fortune in silver and gold.

For this energetic young man of twenty-one, it was an irresistible offer. At last he found the courage to write to his father from London, begging him to let him go. He promised he wouldn't be away for long and told him that his friend Michael Longridge would keep an eye on the factory while he was gone. George refused, but it was no use – Robert had made up his mind to go. But from a letter to his stepmother you can tell the decision had not been an easy one.

> "My dear Mother,
>
> All the temptations you can put before me and all the charms of home will prove fruitless in freeing me from the engagements I have now entered into. Glad would I have been to join my father in his undertakings at Liverpool and I do not even now despair of taking the chief part of his work on myself in a year or two when I return.
>
> This week in London has been one of the most miserable in my short experience. I leave this day at half past 3 o'clock for Liverpool, the ship is said to sail from there in a day or two.
>
> In great haste, *Robert*."

When he got to Liverpool Robert felt even more ashamed about setting off on his adventure and leaving George with so much work. He had told everyone he would be away for about a year, but in fact his South American contract was for three years. His father came to Liverpool to see him off and Robert didn't dare tell him.

They spent three days together being wined and dined by George's rich new Liverpool friends. Robert drank champagne at these farewell parties and talked about his plans for the silver mines and about the riches of South America, with great excitement. Yet a farewell letter to his stepmother shows just how mixed up were his feelings at leaving.

> "My dear Mother,
>
> I seized hold of my pen after the arrival of my dear father to let you know all is well and safe and we are as happy as can be expected considering that I have to part with him.
>
> I scarcely dare think of my father's engagements, indeed if I was to, it would only render me unfit for my own undertaking.

Many will say that I am wrong but I will never say that; I know the experience I shall gain is worth all the trouble. And let me assure you my dear mother it will always be my dearest desire to return home as soon as possible, and by that time I think it will be almost time to get married. What do you think? Will you choose someone for me while I am gone!

I have not time to write any more. I must run off to the ship for she has already weighed anchor and the full sail is spread.

God help you all. I can write no more.

Believe me my dear mother your most sincere
Robert Stephenson."

Poor Robert – like any young man he was trying to grow up and find his own feet, and now he was full of remorse at having put three long years and half the world between himself and his father!

Once he reached South America, Robert was bitterly disappointed. The silver mines of Santa Ana were not at all what he had been promised. The whole project seemed to be in confusion. There were no roads; the mines were miles away from anywhere. The wrong machinery arrived and was sent to the wrong place. He felt frustrated by the impossible conditions of his work. He was also a little afraid, and more than a little homesick. The Cornish miners he had brought out with him despised him, because he was so young and because he wasn't Cornish, and all too soon they took to drinking and fighting.

One way and another Robert was having a hard time, but he bravely wrote enthusiastic letters home telling his stepmother how beautiful the place was and all about the monkeys and the parrots.

"The parrots are either as large as chickens or as small as sparrows, I will try to bring some home with me. We eat the parrots, they are fine roasted. We have an excellent cook who bakes our bread and cooks parrots in various ways so that we never weary of them. Eating and studying take all my time. Though we have an abundance of game around us, we have very few games of amusement!

The Rocket Builder 39

You mention my father intended to write to me from Liverpool, but I have not heard from him. I hope the Liverpool line of road has got through parliament. Let me know everything about home, the factory & etc. it relieves me vastly."

When his stepmother sent him the newspaper reports of the grand opening day of the Stockton–Darlington Railroad which he and his father had planned with such excitement, Robert was quite overcome with homesickness:

"... the description of the scene took me there in a moment. I could imagine the coaches, gigs and people running about in all directions, everyone with a gay countenance. I need not tell you how delighted I would have been to have witnessed the Engine."

Above: Seal of the Stockton and Darlington Railway; *right*: opening of the railway on September 27 1825; *below*: model of *Locomotion*, the engine which hauled the first public railway train

At Darlington George Stephenson had proudly stood on the driving platform of the new steam-engine called *Locomotion* which had been built at the Newcastle factory. A crowd of twelve thousand people shouted with delight and astonishment as he got steam up. Brass bands played and everyone cheered when *Locomotion* moved off with six hundred people clinging to the train of coal-wagons for the long ride to Stockton. The opening day of the first public railway in the world was September 27 1825.

Locomotion, that famous old engine, used to be kept on display on the platform of Darlington Station. You can still see it there today, but now there is a new station and the old platforms have been made into a Railway Museum.

Although Robert had missed that historic day at Darlington, it wasn't all cheerful success at home. George had taken on much too much railway work to do any of it properly, and he was beginning to lose his friends. Some of them wrote to Robert, urging him to come home. Michael Longridge wrote:

"Robert! my faith in engineers is wonderfully shaken. I hope when you return your accuracy will redeem their character. I feel anxious for your return, and I think you will find your father and your friend considerably older than when you left."

The Rocket Builder 41

And from Edward Pease, a sterner note:

> "I can assure you that your business in Newcastle as well as your
> father's engineering have suffered very much from thy absence
> and unless thou soon return, the former will be given up as
> Mr. Longridge is not able to give it that attention it requires and
> what is done is not done with credit to the house."

It was Edward Pease's money that kept the factory going, and
Robert became more and more worried. After getting a long and
loving letter from his father, he wrote home:

> "My dear Mother,
>   My father's letter was an affectionate one and when he spoke of
> his head going grey and finding himself descending the hill of life,
> I could not refrain from giving way to feelings which prevented
> me from reading on. Some, had they seen me, would perhaps have
> called me childish but they are unaquainted with the love and
> affection due to attentive parents, which in me seems to have
> become more and more acute as the distance and the period of my
> absence have increased. You say you long to see me in England,
> believe me, my dear Mother the feeling is mutual, though I am
> almost obliged to stay here till I receive consent from the
> Directors to leave."

It took some time for Robert to get permission to leave, but in the
spring of 1827 he set off for the long journey home. He was to get
a ship from Carthegena in South America to New York and then
another to Liverpool. He wanted to visit Panama on the way, to
explore the possibilities of making a canal across the isthmus.
"How it would influence commerce in every quarter of the world"
he told everyone. But he didn't get to Panama because the place
was in the grip of yellow fever, that dread disease which was to
delay the building of the Panama Canal for another ninety years.
    While he was waiting for his ship to New York, he discovered
to his surprise that his father's old hero, the great Cornish
inventor Trevithick, was staying in the same small hotel. He had
been in Panama with the forlorn hope of building a canal, but he

had fallen sick and lost all his belongings. He was ill, half-starved and penniless when Robert found him and paid his fare home.

Robert had his own adventures on his homeward journey. A shipwreck just outside New York harbour, a stay in New York where he found the people "to have the outward manners of the English but on closer investigation a characteristic impudence, nothing short of disgusting". He went on a long trek into Canada to see the Niagara Falls and from there to Montreal where he bought clothes "of a gentlemanly kind" and threw himself into a mad round of parties and dances for a few days.

It was November before he got back to Liverpool to find that his father really was white-haired and bowed down with worry. Some of the locomotives on the Darlington railway hadn't always been reliable, and there was talk of replacing them with horses. There had been a long political enquiry about the new Liverpool and Manchester railway, a railway George hoped would make use of Stephenson engines. He believed that steam-trains on rails would prove safer than horse-drawn coaches on roads; and he was right.

But in London in front of the smooth-talking parliamentary lawyers, George's brusque Northern manners had been a handicap. He hadn't been able to find the right words to argue with them when they objected to the noise of steam-trains, the black smoke, and the dangerous speed. Nor had he the experience to put before them a well-drawn-up set of engineering plans. Robert's better training would help in the presentation of his father's plans, and Robert was the person to design a simpler and more reliable steam-engine. There was no time to lose if the Stephensons' railway business was to be saved. Robert went back to Newcastle at once to start work again among the skilled mechanics of the Forth Street factory, to do the work he knew best, the designing of engines.

"Rely upon it" he wrote to Michael Longridge, "locomotives shall not be given up. I will fight for them until the last."

He started to plan a different style of steam-engine, working

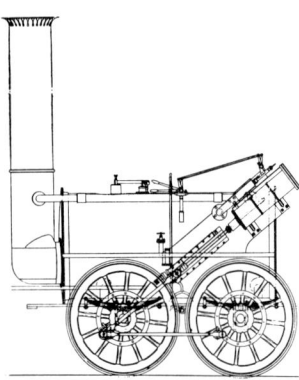

The *Lancashire Witch*

out new designs for the pistons to stop them damaging the track, fitting springs to make the ride smoother, and switching from coal to coke to cut down the smoke.

When it was ready he sent it away for secret trials at a small colliery railway at Bolton in Lancashire. It worked so efficiently and quietly that it was given the name *Lancashire Witch*. Robert was pleased with it and his father was delighted. "An engine for all the engineers in the world to look at," said George, proudly.

And indeed engineers from all over the world came to see the *Lancashire Witch* and placed orders at the factory for similar locomotives to be built for them.

Now that his business was secure again, Robert felt he could ask his London girl-friend, Fanny Sanderson, to marry him and come to live in Newcastle. They found a small house at 5, Greenfield Place not very far away from the factory. Straight after the wedding they went to live there and Fanny brought with her a new-fashioned drawing-room sofa, her piano and her pictures. Today Greenfield Place is an elegant little backwater in a shabby part of central Newcastle, but then it was on the outskirts of the city.

They had little time for honeymooning – Robert was already in the middle of designing yet another new engine, probably the most famous engine ever designed by anyone – *Rocket*. The directors of the Liverpool and Manchester railway were still afraid to use steam-locomotives. They had the idea that they might somehow get loose and run through the streets of Manchester like so many wild beasts. Everyone needed to be convinced that locomotives were reliable and safe, so George persuaded the Directors to stage a public locomotive trial, a sort of competition, with a £500 prize for the most reliable engine. And *Rocket* was to be the Stephensons' entry.

Robert used the successful ideas from *Lancashire Witch* and added a more sophisticated boiler and fire-box, and within five months the sparkling new engine from Newcastle stood waiting to compete with three others: *Novelty*, entered by John Braithwaite

The *Rocket* under steam —
victor of Rainhill and the
most famous of all steam-
engines

and John Ericsson, which came from London; *Sans Pareil*,
Timothy Hackworth's entry from Darlington; and *Perseverance*,
made by the Burstalls in Edinburgh.

The trials were to take place on part of the track close to
Rainhill Bridge near Liverpool, where there was a hill to test the
strength of the engines. Forty laps of the course had to be
covered, a distance of seventy miles in all. The public came in
their thousands to watch, and the locomotive trials soon took on
the atmosphere of a race-meeting, with bookmakers taking bets on
which would win.

As everyone knows, Stephenson's *Rocket* was victorious, but
there's a lovely description of what happened behind the scenes
written in a letter by John Dixon, one of George Stephenson's
engineers.

"We have finished the grand experiments on the engines and
George S. or Robert S. has come off triumphant and of course will

Remains of the *Rocket*, which was modified after its initial journey

take hold of the £500 offered by the Company; none of the others being able to come near them. The *Rocket* is by far the best Engine I have ever seen for Blood and Bone united.

Timothy Hackworth has been sadly out of temper ever since he came. He was grobbing on day and night and nothing our men did for him was right. He got many trials but never got half his 70 miles done without stopping. His burns nearly double the quantity of coke that the *Rocket* does and rumbles and roars and rolls about like an Empty Beer Butt on a rough Pavement. It weighs above $4\frac{1}{2}$ tons and is very ugly and the Boiler runs out very much, he had to feed her with more Meal & Malt Sprouts than would fatten a pig. He must own that he has been taught a lesson in humility.

The London Engine called the Novelty was a light one with no chimney upright, it did not stand above 10 ft. high. The water tank is under the Carriage close to the ground and the Boiler bellows, Flues etc are all covered with Copper like a new Tea Urn, all which tended to give her a very parlour-like appearance and when she started up she seemed to dart away like a Greyhound for a bit but on every trial she had some mishap or explosion, so that it was no go.

Burstall from Edinburgh, upset his bringing it to Rainhill & spent a week in pretending to Remedy the injuries till he was last of all to start and a sorrowful start it was; full 6 miles an hour cranking away like an old Wickerwork pair of Panniers on a cantering Cuddy Ass.

The people were in favour of London from appearances, but we showed them. The first thing old George did was to bring a Coach with about 20 people up at a gallop and every day since he has run up and down to let them see what speed we could do up the Rainhill incline."

That was 1829, and within a year the Liverpool and Manchester line was successfully opened, and from then on everyone went mad about railways and steam-engines. The Forth Street factory got an immediate order for four more locomotives, and Robert was offered the job of setting up the first London-based railway, the London to Birmingham line.

It was to be four times as long as any previous railway, so instead of working from one end of the line to the other, Robert organised his workers to build all the way along the line at once. He had twelve hundred men working for him, most of them "navvies", the nickname for navigators – labourers who were used to hard work because they had dug out the canals with pick and shovel. Now they were building railway cuttings, viaducts and tunnels, and laying miles of line. They lived in shanty towns and moved as the work moved.

Robert was continually on the move too. In the London Science Museum is his leather-covered diary for the year 1834 – the year the first spade of earth was turned on the London–Birmingham line:

"*Feb 4*   Sketching plans of bridges . . .
Feb 7   At Primrose hill tunnel . . .
Feb 10   Saw Mrs S. off North by coach.
Feb 12   At Kilsby to discuss the tunnel . . .
Feb 17   Plans & specifications ready to be concluded.

The London to Birmingham line – making the cutting at Tring in Hertfordshire. One of many drawings by J. C. Bourne who followed the progress of the line

Kilsby tunnel – another of Bourne's drawings

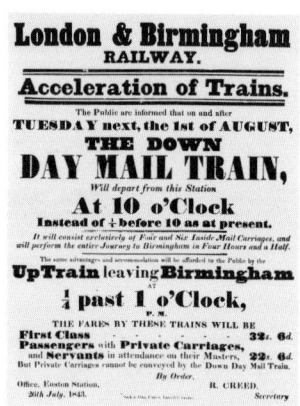

London to Birmingham:
four and a half hours in
1843, one hour forty minutes
in 1981

Mar 19 At Newcastle, went to the factory.

Mar 21 Pease at the factory.

Mar 24 At factory nearly all day.

Mar 28 Mrs S. very unwell and could not leave Newcastle.

Mar 29 To York . . .

Mar 30 to London . . ."

And so on for the rest of the year, to Leicester, to Coventry, Birmingham, Tring, Newcastle and back to London.

All this travelling was done on horseback or by bumpy stage coach. Four years later, he had the satisfaction of driving the first train out of London through the tunnels and across the viaducts all the way to Birmingham. When it was all over, he felt sad; he had worked so hard, for so many years, it felt like coming to the end of writing some great novel.

By now he and Fanny had left their little house in Newcastle and moved to London to live in Hampstead. Robert opened up his own offices in London, but he wrote every week to the factory in Newcastle. He was supplying engines to new railways all over the country – even some to his close friend and rival Isambard Kingdom Brunel for the Great Western Railway.

Sometimes people would ask him who first invented this or that part of the steam-engine. Robert would say: "The railway engine is not the invention of any one man but of a whole nation of mechanics." In fact, British steam-engines were being exported all over the world: to France, Belgium, Italy and America – even to Japan. Robert himself designed a specially decorated train for the Viceroy of Egypt.

But all his success didn't make him very happy. He had no children. He wanted to give his wife presents and, as he was earning a lot of money, he filled Fanny's Hampstead house with paintings. He even went to the College of Heralds and bought a family tree and a coat-of-arms to please her.

But Fanny was continually ill and, six years after they moved to London, she died. Robert was heart-broken. Half the reason for his great success was gone. He sold his house and his pictures; he felt there was only work left to comfort him. His father came on visits to London to try to cheer him up; he played silly pranks in the office, challenging the clerks to Northumbrian wrestling matches, which only succeeded in embarrassing Robert.

Then a sudden and unexpected disaster put a new challenge into Robert Stephenson's life – a railway accident.

Railways had always needed bridges, strong bridges able to withstand the sudden weight of an engine and its load, like the High Level Bridge which Robert had designed for trains to cross the Tyne at Newcastle. In 1846 Robert built another railway bridge at Chester over the River Dee. According to the plans it was to be supported by brick arches, but at the last minute the design was changed to cast-iron girders. Cast-iron seems to be a strong material, but it can be brittle, and in those days the testing of materials wasn't very scientific. Six months after the new bridge was opened, one of the cast-iron girders snapped, plunging the coaches of a train thirteen metres into the river and killing five people.

Disaster at Chester when part of the bridge over the River Dee collapsed

Section of one of the tubes
of the Britannia bridge

It was a terrible disaster, and everyone knew that Robert
Stephenson was already at work on another railway bridge across
the Menai Straits to the island of Anglesey, near Telford's famous
road-bridge. There were rumours that he had lost his nerve and
that the Menai Straits would prove his downfall.

But Robert was determined there would be no more disasters.
He spent a whole year experimenting with models. He couldn't
use a suspension bridge as Telford had done: it wasn't possible for
him to make one stable enough to carry trains. He couldn't use
brick arches, because the Menai Straits had to be left clear for
shipping. He hit on the idea of using iron tubes. Like a rolled-up
sheet of paper, a tube is strong and stiff and light. Huge metal
tubes could cross the Straits and need only a single supporting
pier, which could be placed on a rock that stuck out in the water –
known to seamen as the Britannia rock. So the first tubular bridge
in the world was called the Britannia Bridge.

Actually, Stephenson's tubes were box-shaped and the trains
chuffed their way through the centre on a dark and smoky
journey. It wasn't until later that engineers realised that the
strength of the tube lay in its shape, and trains might just as well
travel in daylight, on top of them. But the idea of the box-tube
was brilliant, a real breakthrough, and Stephenson was justly
proud of his Britannia Bridge.

In fact box-shaped tubes – steel box girders – continue to this
day to be one of the basic structures for building bridges. Even
though we have modern machinery and materials today, the
largest suspension bridges now being built combine the two
principles that Thomas Telford and Robert Stephenson worked
out with such care to solve the problem of crossing the Menai
Straits so long ago.

But it's for railways that Stephenson is always remembered and
three months after the successful opening of the Britannia Bridge,
the platform of Newcastle-upon-Tyne's elegant station was the
setting for a remarkable tribute. Long dinner-tables were laid out
with white tablecloths, glass and silver; and in the background the

Anglesey entrance of the Britannia bridge – a nineteenth-century engraving

station was decorated with huge pictures of railway bridges, garlanded in flowers.

Robert, who was still only in his forties, used this grand occasion to drop a bombshell – he announced that he was retiring from engineering.

Though he wasn't particularly proud of his own success, he wanted everyone to know about his father. So after George died, he helped to write a book about that "self-taught genius".

Robert was made President of the Institution of Civil Engineers; he became an MP, he was even offered a Knighthood. He travelled, and his advice was sought by governments all over the world. He was a millionaire when he died. But he said of himself:

"The Robert Stephenson of Greenfield Place, Newcastle, is the Robert Stephenson I am most proud to think of." That young Robert who worked so hard to build the *Rocket*, which was to start the great age of steam-engines, and make such an enormous change to everyone's way of life across the world.

One of the pillars of the Britannia bridge in 1970, after damage by fire. The bridge has since been repaired

The Rocket Builder 51

Isambard Kingdom Brunel.
This famous early
photograph shows Brunel in
front of the massive anchor
chains of his steam-ship,
The Great Eastern.

ISAMBARD KINGDOM BRUNEL

3

The Little Giant

On Christmas Day in 1857 in the old French Orient Hotel in
Cairo two of Britain's most talented Civil Engineers shared a
family Christmas dinner together. Robert Stephenson, who had
been building a railway in Egypt, had invited his close friends
Mr and Mrs Isambard Kingdom Brunel to take a winter holiday
with him in the mild climate of Egypt. Both men were at the
height of their successful careers; they were in their fifties and had
known each other for twenty-five years. Stephenson was eager to
show the Brunels the awesome relics of Egypt's ancient past, and
Brunel toured the streets of old Cairo riding on the back of a hired
donkey. The joint Christmas holiday was a success.

By the following autumn, within a few days of each other,
Brunel and Stephenson were dead.

Both Brunel and Stephenson were only sons of brilliant and
ambitious fathers. Brunel was three years younger than
Stephenson; his father, Marc Isambard Brunel, was a refugee
from the French Revolution. His mother was English, Sophia
Kingdom.

Marc Brunel was an officer in the French Navy when he first
met sixteen-year-old Sophia, who was in France to learn the
language. They met in Rouen, during the darkest days of the
revolution and fell instantly in love. But within days Marc, who
was openly Royalist, was forced to leave Rouen, flee the country

and escape to America on a false passport. The chances of the couple ever meeting again must have seemed almost impossible. In America Marc Brunel got himself a job as a surveyor and then as City Engineer in New York. It took him six years, but as soon as he had enough money, he left for England to search for Sophia. He found her, and when he married Sophia Kingdom, he married England. Although Britain and France were at war with each other, Marc Brunel offered to work in the British Naval Dockyards, making some special tackle of his own invention.

Isambard Kingdom, their third child and only son, was born in London in 1806. By that time Marc Brunel had a new business, making army boots for the soldiers fighting the armies of Napoleon. In 1815, after the battle of Waterloo, supplies for the army were no longer needed, and like the soldiers who returned from the wars, Marc Brunel was unemployed.

He had a vast quantity of unwanted army-boots and he was in debt. Prison seemed only too possible, but he took comfort in his wife and his children. He was especially proud of young Isambard, who already showed signs of great intelligence. Now, when his father faced bankruptcy and the debtors' prison, he sent Isambard away to France to complete his education. He went to the best technical college in Europe, the *Lycée Henri Quatre* in Paris.

So Isambard was spared the bitter years when Marc Brunel was in prison, where his loving wife Sophia accompanied him. Eventually, the Duke of Wellington raised the money to get the Brunels released, and he went further – he recommended Marc Brunel as an inventive engineer to the new Thames Tunnel Company.

There had been several attempts to make a tunnel under the River Thames – it had been the daydream of engineers for centuries – but each time it had ended in failure, even in disaster. Most tunnels were blasted and hacked through hard rock, but the soil beneath the Thames was soft. Below the water was a layer of soft blue clay mixed with pockets of gravel, and beneath the clay were the treacherous sinking sands in which many an earlier tunnel had come to grief.

The plan Marc Brunel put forward was a bold one. He would build a hundred-metre-long tunnel under the Thames, from Rotherhithe on the south side to Wapping on the north. The tunnel itself would be wide and spacious, a dual-carriageway for horses and carts; at each end it would need a road-way entrance, for the tunnel itself would dip four metres beneath the river-bed.

Marc Brunel was so sure that the directors of the company would accept his plans that he moved home and wife and family to Blackfriars to be near the site. At this point, the seventeen-year-old Isambard returned from France to work in his father's office. He started a diary, which began like this:

"It may be curious at some later date to read the state we are in. I am at this moment penniless. Projects are afoot for the Fowey-Padstow Canal, and the Bermondsey Docks. I am preparing plans for the docks in South London. I am busily engaged on the new Gas Engine, and the project for a canal across Panama. Surely one of these may take place!

I am looking forward to building castles in the air – about steam-boats going at 15 m.p.h., going on a tour of Italy, being the first to go to the West Indies and make a large fortune, building a house for myself etc. etc.

How much more likely it is that all this will come to nothing and that I should pass through life as most people do and gradually forget my castles in the air, live in a small house and at the most keep my gig. On the other side it may be much worse. My father may die, or the Tunnel may fail . . . !"

But despite his son's forebodings, Marc Brunel's plans for the tunnel were accepted. He was appointed Engineer to the new Thames Tunnel project at £1000 a year in the spring of 1825. It was a momentous year for British Engineering: the same year saw the opening of the Stockton–Darlington railway and Telford's bridging of the Menai Straits.

The work on the tunnel started with a ceremony, in the

Thames Tunnel – cross-section showing the shaft and a finished section beneath the river bank

presence of the Prime Minister and the Duke of Wellington. Marc Brunel laid the first brick of the opening shaft, and his son Isambard the second. But the whole scheme depended on Marc Brunel's new invention – the Brunel shield.

This was a huge cast-iron shield weighing eighty tons. Once in place, thirty-six miners could stand on wooden platforms and dig away at the soft Thames clay, while behind them bricklayers used the discarded clay to help line the tunnel walls with bricks. The shield protected the miners and inched forward through the soft wet clay. Most of the work was done with navvies' tools, pick and shovel and wheelbarrow. But the tunnel was lit by gas flares, and pumps worked day and night to clear away excess water. Even so, as the work went on under the river, there were times when the men were working deep in foul-smelling river water which dripped into the tunnel from behind the shield.

The scheme was Marc Brunel's, but his son Isambard became

an enthusiastic site engineer. He worked alongside the miners on the shield, and often spent whole nights in the tunnel watching for any change in the water level. Above, the river rose and fell with the tides, and stormy weather or extra high tides could be heard below in the dark damp tunnel.

Marc Brunel's invention – the shield in action

The whole of Europe was fascinated by the Thames Tunnel project: engineers could talk of nothing else. No one before had made a tunnel beneath a river the size of the Thames. Whether they understood the difficulties of the enterprise or not, important visitors to London begged to be allowed to visit it. Royal princes, Dukes and Ambassadors brought their friends down to see the work. Even well-dressed ladies of fashion joined the sight-seers. The Tunnel Company decided to cash in on this public interest, and began to charge a shilling entrance fee and to advertise Tunnel-watching as a fashionable new London amusement. At one time seven hundred people a day visited the Thames Tunnel.

Marc Brunel tried to stop them. It was enough of a strain being responsible for the lives of his workmen, without the added worry of hundreds of fee-paying visitors. There they were, coming down the entrance staircase and walking through the arches of the tunnel as if they were out for a stroll in Kensington Gardens.

But Marc Brunel and his miners knew that at any moment the tunnelling shield could meet a patch of wet gravel in the Thames clay and that unless straw was stuffed into it as fast as possible, river water would come rushing through.

Isambard was torn between the fun of showing important visitors round and the more urgent work in the shield. One night he actually gave an underground concert to celebrate his twenty-first birthday – the music echoing weirdly through the dark archways – but the next night he was back to his routine, working twenty hours at a stretch on the shield, side by side with miners and up to his knees in mud.

Then, one May evening in 1827, it happened. The visitors had just left the tunnel, and the night shift had started work. In the river above the tide was running full, and strange noises echoed through the tunnel. Suddenly a man at the shield gave a shout: "The river is in." Young Brunel and a hundred and sixty workmen ran for their lives to the entrance shaft, while behind them, glinting in the gas-light, a great black wave of water surged down the tunnel. There was a crash, the lights flared up and went out. But by great good fortune all the men were already up and safe. The last one was rescued by Isambard, who dived back into the dark water with a rope tied round his waist.

Next day in the church of St Mary's, Rotherhithe, Marc Brunel and his wife heard the curate preach a sermon on the flooded tunnel. "It was a just judgment on the presumptuous aspirations of mortal men," he told them.

But one aspiring young man was already planning how to repair the damage: Isambard was out in a borrowed diving bell beneath the river, searching the muddy river-bed for the hole in the tunnel. Once found, it had to be plugged with hundreds of bags of clay mixed with hazel twigs, and that work alone took a month. Meanwhile, the pumps were working overtime to clear some of the water from the tunnel itself. Isambard wrote in his diary:

"What a dream it appears to me! Going down in the diving bell, finding and examining the hole! The novelty of the thing, the excitement of the risk . . . the crowds who came to watch."

He was really quite enjoying himself, and as soon as the muddy water in the tunnel was even halfway clear, he was off taking more risks. With two of the miners he took a punt and travelled as far as he could in the dark towards the shield. They needed to discover if it had been damaged.

"I was quite uncertain what might happen," he wrote, "the hollow rushing of water; the total darkness, the glimmering light of the candles we carried. At the end a dark recess, quite dark, a cavern huge and misshapen with water – a cataract of water coming from it – candles going out."

It took six months to clear the water and repair the huge shield. When work got underway again, the Brunels celebrated in true Victorian fashion with an underground banquet. The evening ended with a little ceremony when the miners presented Isambard Brunel with a pick and shovel. They admired his courage and his willingness to stand beside them at the shield. The tunnelling started again and the visitors came back, in no way put off by the accident.

But two months later, on January 14 1828, the river broke in again. This time six men were drowned and Isambard Brunel was badly hurt. Work on the tunnel was abandoned. Robert Stephenson, just back from South America, was visiting London at the time; he wrote back to his family in Newcastle: "The only news from London is that the digging in the Tunnel under the Thames has stopped." In his diary Isambard Brunel wrote:

"14 January . . . I shan't forget that day in a hurry, it very nearly finished me; but when the danger is over it is rather amusing; to be

Banquet in the tunnel; Isambard is standing with his father in the foreground. In an adjoining arch, the working miners were having their own celebration

Disaster, two months later – the flooded tunnel

nearer the truth it was an excitement, which has always been a luxury to me! When we were obliged to run and I was knocked down, I never expected we should get out. The instant I got my breath, I bolted into the other arch and this saved me . . . All dark, I was anxious for the others. I kept calling them by name to encourage them. While standing there, the effect was grand, very grand. The roar of the rushing water; I cannot compare it to anything, the cannons of war can be nothing to it! At last the water came rushing through the opening and I was obliged to be off.

I have been laid up and quite useless for more than fourteen weeks ever since."

He hated being out of action. All around him other engineers were gaining fame and fortune. The Stephensons were building their railways, Telford had just completed his amazing bridge at Menai, London Bridge was being rebuilt by John and George Rennie, someone else had got the contract for London Docks.

"And I," wrote Brunel, "I have been engaged on the Tunnel which failed, what a recommendation! All my castles in the air, all my fine hopes, crash – gone . . . Well, damn all croaking, it can't be helped."

But the tunnel had not failed. Although in 1828 further working was abandoned, it was eventually finished. It was opened in 1841, and is still in daily use by London Underground trains today.

Yet it was the apparent failure of his father's tunnel which forced Isambard Brunel to find his own way in the world. His character would never let him be depressed for very long. At twenty-two a friend described him as "a dark, nimble young man, with blazing black eyes and a great deal of ready wit". He spent part of his convalescence in Bristol and amused himself climbing along the edge of the steep gorge at Clifton watching the ships on the River Avon below make their way to the sea.

For seventy years the merchants of Bristol had been talking about bridging the Avon Gorge. Now they suddenly announced an open competition for designs for a new iron bridge to cross the

Avon at Clifton. Brunel wasted no time. He set off at once to examine the most famous of all the new iron bridges, Telford's suspension bridge at Menai. He spent four days studying it in detail and decided to send in a design for a similar suspension bridge – though his would need to be even bigger – for Bristol.

Altogether he sent in four different designs, all on the same theme. The Bridge Committee was embarrassed: no less than twenty-two designs for iron bridges arrived and they didn't know how to judge them. So they invited Thomas Telford, the grand old man of the Institution of Civil Engineers to be the judge. He didn't like any of the proposed designs and Brunel's suspension bridge came in for particular criticism. "No single-span suspension bridge," he said, "can be built wider than six hundred feet. I know, I built the one at Menai." Brunel's bridge was to have a span of nine hundred feet. So instead of making a judgment, Telford dismissed all the proposals, and offered a design of his own. It was for a huge bridge, supported by two enormous towers, decorated with gothic belfries.

The men from Bristol were taken aback: it was difficult for them to find any reason to turn down Thomas Telford's distinguished design, yet they weren't sure that they liked it. Brunel wrote them a letter: "Why go to the expense," he asked, "of building Telford's vast towers when the rocks of the gorge are already there?" Extra expense was something that the Bristol committee understood: they invited this bold youngster, who had never built a bridge before in his life, to come and explain his own bridge designs to them in more detail. After the meeting Brunel was crowing with delight. "Of all the wonderful feats I have performed, yesterday I performed the most wonderful." He had persuaded fifteen quarrelling men "on the most ticklish subject of all – taste." They ended by "extravagantly relying on my supreme judgment".

Brunel's bridge plans were agreed and the committee displayed them to the public as an "ornament to Bristol and the wonder of the age". The seventy-three-year-old Telford was certainly not

Clifton suspension bridge –
a view from the air

pleased. But this notorious quarrel with Telford made Brunel's
name, and ever afterwards he referred to Clifton bridge as "my
first child, my darling".

When work started on it, Brunel again caused a sensation by
being suspended over Clifton gorge in a basket and hauled from
one side to the other by ropes. Before very long, money for the
bridge ran out and work stopped: indeed, work on the Clifton
suspension bridge was to stop and start, stop and start again all
through Brunel's life. The bridge was eventually completed five

years after his death by his fellow engineers as their own personal memorial to him. It remains a true ornament to Bristol and one of Brunel's most inspiring works.

But in the 1830s railways were the thing which captured the public imagination. Liverpool, Manchester and Birmingham already had plans for railways to connect their cities with London. If Bristol was to retain its importance, it would need a railway too. And who could advise the merchants of Bristol better than the bright young engineer Brunel, who had offered them the loveliest, and what was more important, the cheapest design for

Telford's elaborate design for a bridge at Clifton which was rejected

Clifton bridge? But Isambard Kingdom Brunel was not prepared to make his name, fame and fortune by being a "cheap" engineer. He told them he was prepared to provide them with plans for a railway to London. He would ensure that a survey and estimate of costs would be completed and ready for their approval within three months. His railway would not be the cheapest; but he could promise them that it would be the *best*. Brunel had never built a railway before, either, but his confidence was so impressive, the committee gave him his head.

He spent the next three months on the survey, riding along the route on horseback by day, and drawing up plans at night. "It is harder work than I like," he complained, in his diary. "I rarely finish under twenty hours a day." But he was aware that all over the country a hundred other eager young engineers were prepared to do the same. He completed his very thorough survey and handed it in on the day he had promised.

It would cost £3,000,000, but it would be the greatest and most beautiful railway in the world. The route he planned would be simple, straight and flat; the valleys would be bridged, he would cut through the hills and tunnel through the rocks. The whole line would be smooth, smooth as a billiard table. Somehow he persuaded the Bristol railway committee to accept his plans. He even persuaded them to change the name of the line. "The Bristol–London Railway" was rather dull – why not call it "The Great Western Railway"? They agreed. They even accepted his estimate. Only a few years before, Thomas Telford had felt ashamed that the construction of the Caledonian Canal had cost £1,000,000. So keen was the competition to get on the railway band-wagon that just ten years later £3,000,000 was thought an acceptable estimate to the careful citizens of Bristol. Of course, there were grumbles. The Duke of Wellington was heard to complain: "All these railways will encourage the lower classes to move about". The Headmaster of Eton, that most famous school, formally objected that the new railway "would bring the evils of London, even the corrupt and revolutionary ideas of France, to the very doors of Eton College". The plans had to get through Parliament and all the objections had to be faced. It was agreed that no railway station would be built within three miles of Eton College – but a few years later, at the personal request of Queen Victoria, a branch line was built to serve Windsor Castle just over the river from Eton. But the most serious objections were to the long tunnel Brunel had planned between Chippenham and Bath under Box hill. It was to go straight through a rocky ridge, and to be nearly two miles long. Memories of his father's uncompleted

Thames tunnel were brought up. "No person would wish to be shut away from daylight for so long. The passengers would be unable to breathe, delicate personages might even die from fear of the dark." It needed all Brunel's tact and charm to get the plans for Box tunnel agreed. By the autumn of 1835, the plans had received parliamentary approval, and work on the railway started in earnest.

That Christmas, Brunel restarted his diary:

Work in progress on Box tunnel

"When I last wrote in this book I was just emerging from obscurity. I had been toiling unprofitably at numerous things. The Railway was still very uncertain. What a change, the Railway is now in progress. I am their Engineer to the finest work in England, a handsome salary, £2,000 a year – and all going smoothly. But, what a fight we have had, and how near defeat, and what a ruinous defeat it would have been. It is like looking back on some fearful pass – but we have succeeded. Bristol is alive and turned bold with this railway. Clifton bridge – my first child, my darling is actually going on, work recommenced last Monday. Glorious!!

The Little Giant 65

And all this at the age of 29, I can hardly believe it. I have taken a house in Duke St – a fine house. I have a fine travelling carriage, I have a cab and a horse, I have a secretary. *I am now some body.* Everything at this moment is sunshine. I don't like it, it can't last. I foresee one thing, this time twelve months I shall be a married man. Will it make me happier."

By June the next year he was married to Mary Horsley. She enjoyed Brunel's fine London house, and his carriage with its liveried footmen and four fine horses. Her brother John was a painter and has left us many of the best portraits of his famous brother-in-law. But Mary Brunel commissioned the fashionable painter Landseer to paint pictures for her drawing-room, and arranged for Mendelssohn, Queen Victoria's favourite composer, to play the piano at her evening parties. They had three children, and Brunel could be an entertaining and loving father.

But he was often away from home and would sometimes say that really his work was his only wife. Certainly he gave his creative energy and careful thought to his wonderful Great Western Railway. He had planned its route to be fast, comfortable and safe. So well did he do it that modern high-speed trains follow the same route as smoothly as Brunel's first trains did a hundred and forty years ago.

He built the railway from both ends, east from Bristol and west from London. At Maidenhead he built a railway bridge across the Thames with two long, flat arches. At the time, everyone said it would collapse – the arches were too flat – but it's still there, and today's much heavier trains thunder across it every day in safety. Between Maidenhead and Reading Brunel hit problems. He had planned a long, deep cutting for the railway at Sonning: it was twenty metres deep and over three kilometres long. The work here almost came to a standstill, so Brunel dismissed the contractor, took off his jacket and got down to work himself, directing his small army of navvies in a sea of mud. Today, the wide straight railway track runs like a secret river through the suburbs of Reading.

He left the real problem, Box tunnel near Bath, to the last. Meanwhile, parts of the line could be opened and starting to earn money. Brunel had his own ideas about the engines he wanted for his railway. He wanted them to be very large, very fast and very spectacular. He ordered six to be made to his own strict specifications. They were spectacular enough, but they were under-powered, unreliable monsters.

Fortunately, at the same time he took on a brilliant young engine-designer, Daniel Gooch. Gooch was only twenty-one when he came to work for Brunel, and with his usual boldness Brunel put him in sole charge of all the GWR engines. It was a choice he never regretted. Gooch had learnt his engineering in Robert Stephenson's factory in Newcastle and he was able to design reliable engines which still had the bright brass funnels and monster wheels that satisfied Brunel's taste for the spectacular.

Gooch and Brunel were in complete agreement about their railway – nothing but the best would do for the GWR. They trusted each other and their friendship was to last a lifetime.

The Sonning cutting near Reading on the line to Bristol – a drawing by J. C. Bourne

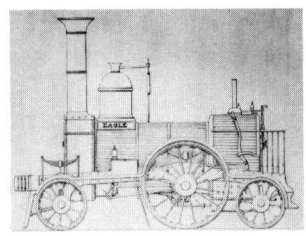

Early GWR engines *above*:
Eagle, below: *Ixion*, Gooch's
improved design of 1839

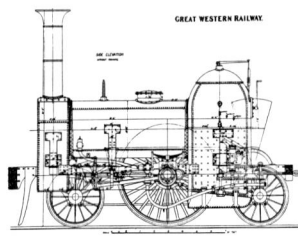

J. C. Bourne's drawing of the
west end of Box tunnel

Gooch planned a whole new railway town for the Great Western at Swindon. The vast engine-works in the centre were surrounded by streets of houses: rows and rows of new homes for the Swindon railway-workers. And even now, though the railway is nothing like as important to Swindon as it once was, you still find there whole families proud to claim that their grandfathers, fathers, aunts and uncles all worked for the Great Western. Under Gooch's management the steam-locomotives made at Swindon set a standard of excellence that for a hundred years was an international legend: the GWR steam-engines with names like the *Iron Duke*, the *Great Britain*, and the last one of all, the *Evening Star*.

While Gooch was busy in Swindon, Brunel was completing the last link in the Bristol–London line. For five years four thousand men dug into the heart of the hill at Box to make the long tunnel. The miners used gunpowder to blast their way through the hard rock, while the navvies carried out the broken stone in buckets. It was known as the hell-hole. There were sudden floods, accidents, uncontrolled explosions; in one year, one hundred and twenty casualties ended up in Bath hospital. The tunnel had been started from both ends, and during the final stages, Brunel took charge. When at last the breakthrough came and Brunel saw for the first time the whole two-mile length of the dead-straight tunnel, he was so delighted he took off his gold signet ring and gave it to his foreman.

The tunnel is said to be so straight that at sunrise, on one day in the year, a long shaft of sunlight fills the tunnel from end to end. With the opening of Box tunnel, the Bristol–London link was complete. In 1841 the first train from London, decorated with garlands and flags, steamed into Temple Meads in Bristol. Temple Meads station was designed by Brunel himself in every detail, from the great hammer beams of its roof to the delicate iron lamp-posts on the platform. The railway had taken six years and more than £6,000,000 to build, twice as much as the estimate.

Nevertheless, it was a great success, and the crowning glory

came six months later, when Queen Victoria announced she
wished to take the train – her first-ever railway journey – from
Windsor to London. A special royal carriage was prepared, its
sofas covered in crimson and white silk. And Daniel Gooch, with
Brunel standing proudly beside him, drove the GWR train that
brought the young Queen to London.

She wrote that day to her Uncle Leopold, King of Belgium:
"We arrived by railroad from Windsor, in half an hour, free from
dust and crowds and heat and *I am quite charmed with it.*"

It was the first of many specially-decorated royal trains that
Queen Victoria continued to use throughout her long life. But
Brunel had set out to bring comfort, even luxury, to all his
travelling passengers, and to do this he had planned his engines,
his trains and his track to be almost twice as wide as any others.
The Stephensons, who had built the first public railways, had
colliery tracks as their starting-point, and their standard gauge
(the width between the lines) came from the width of colliery
trucks, four feet eight inches wide. An extra half-inch had been

added to give passenger trains some leeway on curves.

Brunel insisted that trains and track seven feet wide would be better, safer and faster. Quite independently he built his railway on tracks nearly twice as wide as anyone else's, though even he had had to add an extra quarter-inch for safety. The broad-gauge trains were certainly more comfortable, and Brunel might well have been right – but he was too late. By the time he completed the GWR, nine-tenths of the railways then existing had been built to the Stephensons' standard gauge. There was no way that the standard gauge lines from the North which came into London at Euston could link up with the broad-gauge GWR lines with their terminus at Paddington. One railway line, however beautiful, cannot stand alone in a whole nationwide network of railways.

It caused bitter battles and legal squabbles in and out of Parliament. In the end, Brunel had to give in. His broad-gauge tracks were torn up, and his great engines with their luxurious rolling-stock altered. It cost the GWR a fortune and took forty years to complete – and to this day there is no main-line railway link across London.

But it is easy enough to forgive Brunel for his stubbornness – he put a lot of work and enthusiasm into his engineering projects, and often backed his ideas with his own money. Sometimes he lost.

He wanted his Great Western Railway to go further than Bristol, to go on to Exeter, South Devon and Cornwall. Unfortunately, he got involved in an experimental idea called the "atmospheric railway", and tried it out on the South Devon line. Within six months he realised it was a failure and got rid of it. But having been so wrong with the experiment, he decided to accept no fee for building the line. The railway was built, and the tracks still run hugging the Devon coast-line between the red cliffs and the sea.

With all his work on the GWR and its branch lines, and his continuous attempts to complete Clifton bridge, Brunel had enough work to keep him happy. But his energy and enthusiasm

Approaching Teignmouth in Devon – a broad gauge train on one of the most spectacular stretches of line in Great Britain

was endless. Side by side with his railway work, he was busy designing ships; like his railway trains they were to be bigger, better and more luxurious ships than anyone had built before.

When he had first set out the financial plans for the GWR, one of the company directors sighed, and complained that the distance from London to Bristol was so enormously long. Brunel's immediate reaction was: "Why not make it even longer? From London to Bristol, and from Bristol to New York. The steam-train to take passengers to Bristol, and a steam-ship to take them across the Atlantic to New York? We could always call the ship the *Great Western*, too." His listeners weren't impressed – they thought he was joking. But he wasn't. He went quietly off to Bristol and set up the Bristol Steam-ship Company to build the *Great Western*, a great wooden steam-ship designed by Brunel to cross the Atlantic to New York.

At that time, steam-ships were used only for short journeys round the coast. They used coal to heat their boilers, and it was thought dangerous for them to travel too far away from stocks of coal. But Brunel convinced his fellow ship-builders that if you made the ship large enough, she would have the storage space to

Great Western on her maiden voyage 1838

carry sufficient fuel for any long journey, even to cross the Atlantic. The *Great Western* was to be a paddle-steamer with her wooden hull specially strengthened against Atlantic gales. She was to be large enough to get to New York without refuelling. Her saloon alone was twenty-five metres long – longer than any other steam-ship in the world. She was painted black, and her prow was beautifully decorated with golden dolphins.

Soon other shipping companies got to hear of Brunel's plans, and then everyone wanted to be in on the race to have a regular passenger steam service across the Atlantic. The London Steamship Company hired a fast paddle-steamer, the *Sirius*. She was only half the size of the *Great Western*, but by increasing her coal-bunker space they reckoned they might manage to make the Atlantic crossing before Brunel had got his own ship ready, and spike his guns. It was a race without any rules.

In March 1838, in secret, the little *Sirius* set off. But Brunel had seen her leave. Within three days the *Great Western* was ready and away. Brunel was aboard and thrilled that the maiden voyage of his new ship should be in such a blaze of publicity. The race was on; the enthusiasm spread to the crew, but after two hours of travelling full-steam ahead, smoke was pouring from the boiler-room. Some loose lagging had caught fire, but the damage was slight and could be repaired in a few days.

Unfortunately while Brunel was making his inspection, blinded by the smoke he slipped and fell from the top to the bottom of the engine-room ladder. He was in such pain he was unable to move. But he cared so much about the race that he insisted he should be put ashore, and that as soon as the repair-work was finished, the *Great Western* should continue the race. He was wrapped in a sail and gently rowed ashore.

The *Great Western* was put right, but by now she was eight days behind the *Sirius*, which had had time to refuel at Cork in Ireland. In the Atlantic both ships ran into heavy seas. *Sirius* did reach New York first, but she was only just in time: her fuel bunkers were quite empty. Twenty-four hours later, the huge black *Great Western* arrived in New York harbour. She had lost the race, but she still had over two hundred tons of coal aboard.

Brunel had proved his point that a large steam-ship was a safe, reliable and comfortable way of crossing the ocean. The *Great Western* could carry passengers across the Atlantic in luxury and style, keeping to a regular time-table. In fact, in the next eight years the *Great Western* made sixty-seven crossings, her fastest in twelve and a half days.

But once Brunel had proved that one of his ideas worked, his next ambition was to make something different and better. No sooner had the *Great Western* returned from her second voyage than he was already planning her successor, the *Great Britain* – even bigger and more spectacular, and something entirely new. She was to be the very first of her kind – an *iron* steam-ship powered by new screw-driven propellers. This time everyone thought Brunel had gone too far. Surely, an iron ship that size would sink, or it would run out of coal, or its new-fangled engines would fail; and most surely of all, Brunel and his Steam-ship Company would go bankrupt.

But Brunel was too busy to care about people's doubts. At the same time as he was working out the calculations for the exact size of his new iron ship, he was also struggling with explosives in the tunnel under Box hill. While he was deciding on just the right way to mount the steam-ship's new propellers, he was also designing every detail of the stations up and down his railway line, and even choosing the right colour paint for the smart new GWR engines.

The sheer energy of this little man quite rightly earned him the nickname of the Little Giant.

In 1843, the year after Queen Victoria made her first train journey, Brunel and Daniel Gooch drove another royal train from London to Bristol. Prince Albert had come to launch Brunel's giant iron ship, the *Great Britain*. There was champagne and a banquet, and many toasts were drunk to Bristol and to Brunel – and to the ship itself, the Queen of the waters. This was the last time Bristol was to see its most famous ship in all her glory.

She was too big for Bristol docks and the port authorities weren't going to enlarge them – so she was off to London to be fitted out there. A hundred and twenty-seven years later, after many adventures all round the world, the iron hull of the *Great Britain* was brought back to Bristol, still afloat and still much the same as Brunel had first planned her. She is now a Bristol show-piece, and being repaired, repainted and redecorated as a memorial to Brunel and the Bristol ship-wrights who first built her.

She was an astounding ship, with sixty-four stateroom cabins, her own music-room, a vast saloon. She was one third larger than any other ship afloat at the time. In London, Queen Victoria and crowds of other awestruck Londoners came to admire her size and the lavishness of her passenger accommodation. Specially-woven carpet covered her public rooms. All her china, even her bathroom-ware was specially made with a *Great Britain* design in blue and white.

It was from Liverpool, Bristol's rival port for the Atlantic trade, that the *Great Britain* set out to make her first trip to New York, and thousands of Liverpudlians turned out to wish her good luck. When she arrived in New York, the Americans, who admired Brunel's flair for the new and the extravagant, gave her a royal welcome with flags and fireworks and enthusiastic cheers. She was a great success.

Decorated entrance to a berth on the *Great Britain*

Then, literally overnight, Brunel's success turned to failure. In the autumn of 1846 the *Great Britain* left Liverpool for New York. There was no reason to think that anything would go wrong: the ship had more than proved her sea-worthiness, having already made eight successful Atlantic crossings. But in the dark of that first night at sea, the ship lurched – and the passengers were suddenly jolted out of their red velvet saloon chairs. People screamed with fear, but nothing else happened. The ship wasn't sinking, but she wasn't moving either. In the dark, the great iron ship had quite unaccountably run aground on the wide sands of Dundrum bay on the east coast of Ireland. No one was hurt: even the ship was scarcely damaged. But there she sat in the sand, like a beached whale. It was an absurd sight.

Next morning the local Irish peasants crossed the sands in their farm-carts to examine the strange iron monster that fate had landed on their shores. They made a small fortune rescuing the passengers. But who was going to rescue the ship? There were great arguments about how she got there. Were the new charts her captain used wrong? Or had the iron hull distorted the compass readings? However it happened, there she lay helpless, with winter and its gales approaching.

The Bristol Steam-ship Company went bankrupt. But for Brunel, losing money was not half as bad as losing his beautiful ship. While his bankers were arguing about who owed whom how much money, the *Great Britain* lay stranded. Brunel went to examine her himself: and what he saw made him furious.

The Little Giant 75

Great Britain stranded off the coast of Ireland

"The finest ship in the world has been left lying about like a useless saucepan," he wrote home. "It is positively cruel. She is beautiful, perfect; just a little bruised. But she is being left to go to pieces, as if her own parents meant her to die there."

If no one else was going to do anything to save his ship, Brunel was. With the help of the local farmers, he built huge cushions out of branches and stakes and chains, and then lashed them to her sides. They might protect her from the winter storms. In fact, it worked so well that the *Great Britain* was floated off on the spring tides and brought back to Liverpool for repairs. But Brunel had been forced to sell her. Though it was by no means the end of his iron ship. During the Crimean War she was used as a troop-ship, taking British soldiers out to the Black Sea.

In the Crimean War, the English and French armies were

fighting side by side, and it was then that for the first time some serious plans were produced for a Channel tunnel. It was to be a dual railway-track inside an iron tube, and halfway across there was to be a floating island where the trains might stop and the passengers get out, take the air and admire the view. At night, the island would be lit up like a fun-fair. Both Brunel and Robert Stephenson became interested and signed letters of support. But politics intervened, and the Channel tunnel is still just an idea to this day.

But Brunel already had three new projects he wanted to complete.

One was to take the GWR across the Tamar Estuary into Cornwall at Saltash, by building the largest railway bridge he had yet attempted.

The second was to build a fine new station at Paddington, to be entirely made from iron and glass, and next to it a luxury Railway Hotel. The care and comfort of his passengers was very important to Brunel. He was the first person to complain about railway coffee. He wrote to Gooch about the coffee-rooms on Swindon station: "It's not just inferior coffee, I don't believe they buy coffee at all. I avoid taking anything there, I have ceased making complaints about it." For the hotel at Paddington station, he would choose the chef himself and even approve the menus, if necessary.

The third plan was for another ship. On the back of one of the designs he drew for the delicate ironwork of Paddington station, there's a sketch of a ship with six masts and five funnels. This ship was to be the last and grandest of all Brunel's schemes. There was a gold-rush to Australia in the 1850s, and Brunel's first iron ship, the *Great Britain*, was already making successful journeys taking six hundred and fifty passengers at a time out to Australia. Brunel was planning another iron ship, the *Great Eastern*. A ship of 2,400 tons, 300 metres long, which would take four thousand passengers halfway round the world to Australia. No one was to plan anything as large as that for another fifty years.

She was built in London, at Blackwall, where Brunel could keep an eye on her progress. It took four long and troubled years to build her. And when at last the day arrived for the launch, there was even more trouble. The ship was too big to be launched in the conventional way: she needed to be sent sideways into the river. Brunel knew it was going to be a tricky operation, needing all his most skilled men's attention and his own complete concentration. He needed absolute silence if he was going to succeed.

His friends, Daniel Gooch and Robert Stephenson, went with him to Blackwall to give him moral support. But when they got there, they found the dock-side crowded with sightseers. Tickets had been sold for the best seats, and the noise was deafening. Brunel was in two minds whether to continue. When he started the delicate operation of moving the giant ship, she moved only a few feet and settled into the Thames mud. The crowds booed and jeered, and in the confusion one of the workmen slipped and fell to his death. Brunel stopped the launch at once, and went home.

Next day, the newspapers attacked him:

> "Will this be another monument to Mr Brunel's vanity? We seem to be a little unfortunate in our grandiose schemes of late. If great engineering consists in effecting huge monuments of folly at enormous cost, then is Mr Brunel surely the greatest engineer."

There's nothing the press likes more than to kick a hero who slips in the mud. Robert Stephenson wrote at once to Brunel, assuring him he would stand by him always, and would help him in any way he could.

A few weeks later, without letting anyone know, they inched the *Great Eastern* quietly into the water. They used special hydraulic rams to move her, similar to the ones Stephenson had used when he put up the iron tubes of the Britannia Bridge. It wasn't easy, and Brunel worked for sixty hours without sleep. The *Great Eastern* was successfully launched, but it seemed that she might get no further. Brunel had spent all his money, and his

The biggest of Brunel's ships, the *Great Eastern*, 22,500 tons

health was broken. The press went into the attack again, predicting that this monster ship would stay forever in the River Thames and end its days as a giant fun-fair.

But at least at Saltash things were going well. The bridge to Cornwall was 730 metres from end to end. Two arched spans made from iron tubes were to rest high on a massive central pier and carry a single track suspended on an iron deck.

Each span had to be floated out into position at high tide, and then slowly raised up by huge hydraulic jacks. This was always the trickiest operation, so on that day, as the first span was floated out into position, Brunel in his top hat stood alone high up on the massive centre pier. From there the little man shouted his instructions, while the work proceeded in strict silence. When it was all done, and both bridge spans were in place, from away on the Cornish side of the water came the sound of a brass band playing *See the Conquering Hero Comes*.

The Little Giant 79

Rejoicing all around as one of the giant spans of Saltash bridge is hoisted into place

But by now the conquering hero was a sick man. He took his first holiday in years and, with his wife, joined his friend Stephenson in Egypt.

While he was away, Queen Victoria and Prince Albert paid an official visit to the *Great Eastern* at Blackwall. No doubt this visit helped to encourage people to invest in her future, because from then on, work re-started on the ship.

But Brunel was ill. He came back once more to see Saltash Railway Bridge at work, and Daniel Gooch arranged for his sick-bed to be placed on a flat platform truck. Gently a giant GWR locomotive pulled the little figure of Brunel across his magnificent bridge for the first and last time.

A few days later, he and his wife went to look over the *Great Eastern* – the ship of which he had said: "I never embarked on any one project to which I have devoted so much time, thought and labour, and on the success of which I have staked so much reputation." At last the ship had been fitted out. All the red velvet and silk furnishings were in place. The great saloon was decorated with palm trees and mirrors and great glass chandeliers. Mr and Mrs Brunel were delighted to see her completed – the ship of his dreams. They had tickets to sail on her first voyage.

But when the *Great Eastern* set sail, Brunel was at home in bed, ill. Days later they brought the bad news to him there. Two days out, there had been an accident in the boiler-room. By a mistake, one of the steam-cocks had been left in the wrong position: a silly mistake, with devastating effects. There had been an explosion, the deck above had split open: bits of glass chandeliers and gilded mirrors fell like rain. Down in the boiler-room, in a rush of scalding steam, six men were burnt to death.

The ship survived: she was afloat; she could be easily repaired, but for Brunel the news of the accident was the final blow: within a week, he was dead.

Daniel Gooch, who had known him better than anyone else, wrote his epitaph:

"By his death the greatest of England's Engineers was lost; the greatest originality of thought and power of execution. He was bold in his plans, but right. The commercial world thought him extravagant; but although he was so, great things are not done by those who sit down and count the cost of every thought and act."

Isambard Kingdom Brunel truly earned his title "the Little Giant".

Spaciousness and luxury afloat on the *Great Eastern:* the Grand Saloon

82 *Breakthrough*

4

The Ditch in the Desert

The steam-ships that Brunel built in England were a true breakthrough. Unlike traditional sailing-ships, they weren't at the mercy of the winds, and Brunel proved they could be built large and strong enough to cross the Atlantic Ocean. They could also steam south from Britain to the tip of Africa, cross the Indian Ocean and reach the Far East, Australia and beyond.

Meanwhile another great pioneer, Ferdinand de Lesseps, was planning to cut by half the time it took to make that long journey to the East. De Lesseps was building the longest ship-canal in history straight through the Egyptian desert, to link the Mediterranean Sea with the Red Sea.

Ferdinand de Lesseps – that's a name you don't hear much about today, but a hundred years ago he was world famous. He was a Frenchman, much talked about in England, who did the thing everyone said was impossible – he built the Suez Canal.

Those were the days, so the saying went, when the sun never set on the British Empire. In the east, it stretched from India to Malaya, Singapore and Australia, and when the Suez Canal was completed, British steam-ships poured through it taking civil servants, troops, planters, teachers and their families to new lives in its far distant corners. Back through the canal came ships carrying cotton, rice, timber, spices and all the rich raw materials of the East to the industrial cities of Europe. Most of the ships

Opposite: Ferdinand de Lesseps in 1853

using the canal were British. By making the canal, de Lesseps had created a quick sea-connection between Victorian England and her Empire, and it became known as the "Gateway to India".

For a hundred years, while the British Empire existed, the Suez Canal was almost as important to Britain as the English Channel. Many of the men who worked in the canal, piloting ships from Port Said through to Suez, were British; much of the money the canal earned came to Britain; British troops guarded its banks. Every year, thousands of British ships steamed through it. You might have thought the canal actually belonged to Britain. During two wars this century, the British army fought to defend Egypt because of the Suez Canal.

But the strange thing is, when Ferdinand de Lesseps first talked about his plans, the British tried everything possible to stop the canal being built. For fifteen years the powerful government of England fought the ideas of one enthusiastic Frenchman, and the reason they didn't trust him was simply because he was French. In the Middle East, France and Britain were rivals. Since the days earlier that century when Napoleon's Army had occupied Egypt for three short years, the British Foreign Office watched warily for any new French moves in Egypt. But Ferdinand de Lesseps didn't want his canal to be French. His dream was to build an *international* sea-route, a waterway "guaranteed for the use of all nations and the universal freedom of the seas". To make his dream come true, he had to fight very hard.

Ferdinand de Lesseps came from a family who looked at the world through international eyes. His father, his uncle and his grandfather were in the diplomatic service and worked as French consuls in many different countries. He grew up in France at exactly the same time as Robert Stephenson and Isambard Brunel were growing up in England. In fact when young Brunel was sent over to continue his education in Paris, he went to the same school as de Lesseps. They were there at the same time: perhaps they knew each other. Certainly in later life de Lesseps knew Robert Stephenson, and the two of them very nearly came to blows.

The romance of the past –
a view of Egyptian travel
published in 1855

While he was still at school, Ferdinand decided to follow the
family career and study languages, hoping to become a diplomat.
What attracted him was the glamour of travel and the stories of
far distant places.

In 1832, he was given his first proper appointment and sent as
French Vice-Consul to Egypt. When he arrived in Alexandria,
young Ferdinand de Lesseps caused quite a stir. He was elegant,
witty and a superb horseman. He had the reputation of being a
good dancer who loved the company of women, and they found
him equally charming.

He was officially welcomed in Cairo by the Viceroy, Mahomet
Ali, who remembered his father. Like many Eastern rulers in
those days, Mahomet Ali had a harem full of wives, he also had no
less than eighty-four children. But one of his sons, called Saïd,
was quite a problem for his father. He was an intelligent and
sympathetic child, but much too fat. As a favour, the Viceroy
asked the young French consul to take Saïd under his wing.
Perhaps he could teach him to ride, give him a few fencing lessons
and even make a sportsman of him. Saïd never became much of a
sportsman, nor did he get any thinner! Whenever his father put
him on a strict diet, he nipped round to Ferdinand's house to beg
the cook for a plate of spaghetti. In fact, whenever his father was

Mahomet Ali in about 1840
– from an early type of
photograph

The Ditch in the Desert 85

angry with him, Saïd went running there to hide. Ferdinand didn't mind, he was quite fond of the boy.

But meanwhile de Lesseps was busy finding out about Egypt. He learnt Arabic, and studied Islamic law, and discovered all he could about ancient Egypt. Once, while reading about the country's recent history, he discovered that Napoleon had searched for traces of an ancient canal, a waterway that had once connected the Nile with the Red Sea.

De Lesseps began to have his own dream of digging a canal; a wide, deep, modern canal like a branch of the sea, cutting through the Isthmus of Suez. He had visions of large ships sailing straight across the sands of the Egyptian desert in a canal which would link Europe with the East.

There was already a desert road which joined Cairo to the Red Sea. It was used by British travellers going to India – they called it the "overland route". But that was rather a grand name for what was really a dangerous, dusty desert path. Thomas Waghorn, a lieutenant from the Indian Army, had discovered the route and spent years trying to make it safer and more comfortable.

Ships of the P & O line brought passengers and mail from England to Alexandria on the Mediterranean coast, and other ships called at the little Red Sea port of Suez bound for Bombay. But over a hundred miles of hot, lonely desert separated the two ports. Waghorn arranged a series of rest-houses along his route, with fresh horses for the travellers and little open carriages for the ladies brave enough to undertake the journey. He even hired armed guards to protect the convoy from bandits. But when the travellers reached Suez they sometimes had to wait weeks for a steamer to call. It was not much of a route, but de Lesseps admired Thomas Waghorn for making the road possible at all. Many years later, when the Suez Canal was open, he built a monument to Waghorn saying: "He led the way and we followed."

PANORAMA DU CANAL DE SUEZ

MER ROUGE

SUEZ

PETITS LACS

GRANDS LACS AMERS

LAC TIMSAH

ISMAILIA

El-Ferdane

LACS BALLAH

KANTARA

LAC MENZALEH

PORT-SAID

MER MEDITERRANEE

Route of the Suez Canal.
This panoramic map has the
Red Sea at the top and the
Mediterranean below

The Ditch in the Desert 87

That was all to come much later. For the present, the young French diplomat had to face one of the many duties of being a good consul. In 1836 there was an outbreak of cholera which quickly turned into an epidemic. Suddenly and terrifyingly, people were dying in thousands all around him. The epidemic raged for two years, and during that time a third of the inhabitants of Cairo died.

Ferdinand turned the French Consulate buildings into a hospital and worked himself among the sick and dying patients. It was an heroic effort, and even his British colleagues admired his courage. They made an official commendation: "The French Government is to be congratulated in having men of the stamp of M. de Lesseps in her foreign service. We have never seen anyone so young represent his country more creditably."

His term as consul in Egypt ended soon after; he had gained an insight into the lives of ordinary Egyptians, and had made friends with one of the Viceroy's sons, Saïd. But his most enduring thoughts were concentrated on the almost impossible idea of recreating that ancient canal to link the two seas.

As soon as he returned to France he got married, and then set off with his new bride, Agathe, to his next diplomatic post in Spain, as consul in the war-torn city of Barcelona. Once again, he allowed the Consulate to become a refuge for anyone in need.

"Everyone loves and admires him," his wife wrote home. "Whenever he appears, they say 'The Consul of France', and make way for him."

His career as a diplomat reached its height when, in 1848, at the age of forty-three, he was made Ambassador to the court of the young Queen of Spain. Among the ladies at the court in Madrid his social graces and charming manners had quite an effect. One of the Queen's closest friends was a beautiful red-haired girl of twenty-two, Eugénie de Montijo, whose family was related to the de Lesseps. Eugénie was rich and lively and very popular, and she and Ferdinand became firm friends. Indeed she was in some danger of having her head turned by a long list of devoted

Eugénie de Montijo – who became de Lesseps' "guardian angel"

admirers, but Eugénie's mother soon whisked her off to Paris to find a suitable husband.

A year later, Ferdinand de Lesseps was also back in Paris: the new French Emperor, Napoleon III, wanted to send a special envoy to Rome. He chose de Lesseps to tackle a difficult, almost impossible, task, to negotiate peace with the Pope. Meanwhile, without his knowledge, the French Army had been sent to Italy to make war. In the middle of this, de Lesseps was recalled and his mission was declared a failure. It seemed he had been used as a scapegoat. Ferdinand was disgusted by the whole affair, and immediately resigned. It was a sad end to his bright career and to his family's long connection with the French Diplomatic Service. Worse still, without his government salary he was penniless.

He retired to his wife's farm in the country and busied himself learning to be a farmer. There he found that his thoughts kept returning to Egypt and to his dreams of building a canal. He turned to books on geology and engineering, to study canal-building. Different engineers had played around with the idea of an Egyptian canal. Even the great Robert Stephenson had gone to Egypt to look at a possible route, but then he was commissioned to build a railway from Alexandria to Suez instead – an idea he vastly preferred. So the Egyptian State Railway was begun. It took eight years to build and cost several million pounds, and the elaborate royal train Stephenson designed specially for the Viceroy helped put up the cost.

But Ferdinand de Lesseps saw in his mind's eye something altogether larger than a couple of hundred British travellers crossing the desert in a train on their way to India. He imagined a canal that would carry the world's trade in great steam-ships travelling unhindered from sea to sea.

Mahomet Ali was dead, and Ferdinand had heard there was a new Viceroy in Egypt. He wondered whether it would be worth putting his plans before him. He wrote for advice to an old friend in Cairo:

It is now three years since I resigned from the Foreign Office. Since that time I have been studying a project that has been in my mind ever since I left Egypt, twenty years ago. I admit my scheme is still in the clouds – I refer to a canal cutting through the Isthmus of Suez. It is regarded as impossible – the obstacles insurmountable, I am the only person who believes it possible. I send you a very confidential document, the results of my studies. Would the present Viceroy understand the great benefits of my scheme? Would he be the man to help in carrying it out? Let me know your opinion.

He got an immediate reply. He was told to forget the whole idea. The new Viceroy, Abbas Pasha, was quite definitely not the man to assist any such schemes. He disliked all western ideas, ideas for modernisation, and particularly ideas which came from France.

De Lesseps had to be content with that. He didn't like retirement, but he enjoyed country life. He had three sons and already they were at his old school in Paris. Then in January 1853 all France was suddenly agog with an astonishing piece of news. The Emperor was in love with a rich and beautiful red-haired Spanish girl called Eugénie, and intended to marry her almost at once. De Lesseps' young cousin Eugénie – the girl who had caused such a stir at the Spanish court – was unexpectedly to become Empress of France.

It was an elaborate ceremony, and when Eugénie appeared in her wedding dress in the Cathedral of Notre Dame to be crowned Empress of France, the Paris public were impressed by her beauty and dignity. The de Lesseps family had scarcely recovered from the excitement of Eugénie's State Wedding, when news came from Paris that one of their sons was ill. It was scarlet fever, and Agathe hurried to Paris to nurse the boy herself.

A few days later, she too had caught the fever, and within a week they were both dead. Ferdinand was utterly heartbroken. He left the other two boys, Charles and Victor, in Paris, in the care of his wife's mother, and returned home alone to the farm. He was in a black mood of depression: his career, his

Wedding portrait of Napoleon III and Empress Eugénie

marriage, even his dreams for the future, were all finished.

Then, out of the blue, he received a letter from Egypt. Abbas Pasha was dead: there was a new Viceroy in his place, who was none other than the fat, jolly boy who used to come to eat his spaghetti – Saïd.

Here was a great new chance – but he must take care not to spoil it. He would go back to Egypt, just to be able to congratulate his old friend. But any talk about his canal must wait for exactly the right moment. He needed to choose carefully the best way to approach Saïd. After all, it was twenty years since they had last met, and twenty years is a long time.

But Saïd Pasha hadn't forgotten his childhood friend, M. de Lesseps. He was given a royal welcome in Egypt and invited to stay at the Viceroy's summer palace. He wrote an almost fairy-tale description of his splendid welcome for his family to read:

> My pavilion is in the middle of a lovely garden, among avenues of flowering trees. The air is perfumed with jasmine. On the first floor there is a very bright drawing-room with four rich divans and four large windows. Leading out of it is the bedroom, with a very elaborate bed, the hangings are of handsome yellow silk embroidered with red and a gold fringe. The dressing-room has rosewood and marble furniture, the washing utensils are in silver, the soft towels being all embroidered with gold.
>
> I go to pay an early visit to the Viceroy, we recline on an easy divan in a gallery overlooking the garden. The servants bring basins and ewers of silver and then sprinkle us with rose-water before bringing food to us on a salver. The Viceroy was in the best of humours, and we spent a couple of hours discussing all subjects.

But not a word did de Lesseps say at this first meeting about his plans for a canal. He wanted to be quite sure that the Viceroy's close advisers approved of him. Most of them knew little about him or of his past friendship with their new ruler.

But Saïd was delighted to meet his old friend again, and he invited de Lesseps to go with him into the desert to watch the

SAÏD-PACHA,
VICE-ROI D'ÉGYPTE.

Saïd Pasha, Viceroy of Egypt in 1854, after whom Port Said was named

army manoeuvres. For ten days they were constantly together. They would go out from the camp on expeditions, to go shooting. Ferdinand did the shooting while Saïd, who was fatter than ever, sat in his royal carriage and watched. One morning Ferdinand woke early. It was dawn. He walked to the edge of the camp.

> To my right the east was already bright, to my left the west remained sombre and dull. Suddenly I saw a rainbow of the most brilliant colours stretching from west to east. I felt my heart jump. Was this an omen that today would see the success of my project, my practical plan to unite the West and the East?

Saïd had given de Lesseps a fine Arab stallion, and he and his Army generals gathered round to see how the Frenchman would manage this spirited horse. Ferdinand felt sure that this was some sort of test. Everyone was watching him, so on the spur of the moment he set his horse to jump the high wall that surrounded the camp, and with a wave, galloped off into the desert.

When he returned, the Egyptian generals came to call on him in his tent. He had been accepted. That evening, confident that the right moment had come, Ferdinand told Saïd all about his plans for the great canal. By morning the matter was settled. The Viceroy had signed a detailed agreement, giving his friend "the sole right to establish a Universal Company to construct a canal, a passage for large vessels, through the Isthmus of Suez."

De Lesseps had won his first battle, now he had to win the approval of the rest of the world and find the money for his Universal Company. He was not a business man, he was not even an engineer, he was only a man on fire with enthusiasm for an idea. It would be possible for him to employ men with engineering expertise, possible even to find financial advisers, but he alone would have to persuade the governments of Europe to get together and back his international scheme.

It was British approval and British support he needed most. For once, Britain and France were openly friends, since their armies were allies, fighting Russia in the Crimea. Queen Victoria had been most welcoming when the French Emperor brought his new bride, the Empress Eugénie, on a state visit to England in 1854. With some confidence, de Lesseps set off to try his luck in London. He wrote to his family:

I have set my course to accomplish this matter without any personal gain. I am confident I shall be able to pilot my ship into the port we may well call Saïd: for the name of the Viceroy also means "happy" in Arabic. One thing that is happy, is that my actions are entirely my own. I am not accountable to, nor can I be disowned by, any Government.

In England everything started off very well. Queen Victoria gave him an audience, and Prince Albert, who was always interested in new engineering projects, wished him good luck. But the newspapers and the Prime Minister Lord Palmerston were against the plan. De Lesseps had a long interview with Palmerston and tried his hardest to demonstrate how much Britain had to gain from the very existence of a canal and how much research he had done on the engineering problems.

Palmerston was not impressed. "It is," he said, "absolutely against British interests. And anyway the plan itself is quite impractical. All the engineers in Europe can say what they like, but nothing will change my mind, not a jot. I will oppose your canal, M. de Lesseps, to the very end."

For the next five years Ferdinand de Lesseps toured the industrial cities of England, trying to persuade the British people that his canal would be to their advantage.

The East India Company and the shipping line P & O were already his powerful allies. Eventually, his friends asked questions in Parliament. But Palmerston had not changed his mind about the Suez Canal – "not a jot". He suspected de Lesseps of some political motive to promote French interests in Egypt. He couldn't believe that this earnest Frenchman was merely "in love" with an idea. In the House of Commons, Palmerston directly accused de Lesseps of trying to induce Englishmen to part with their money. "I can only express my surprise," he declared, "that M. Ferdinand de Lesseps should have reckoned so much on the credulity of English capitalists as to think he would succeed in obtaining English money for a scheme which is in every way so adverse to British interests. A scheme launched, I believe, as a rival to the new Egyptian railway. My advice to the English is to have nothing to do with M. de Lesseps."

Robert Stephenson, the most eminent engineer in Europe, was now an MP. He was building that railway in Egypt. He added his advice to the Prime Minister's. "To dig a ditch through the desert," he said, "where there is no water and no means of securing supplies, I will not say it is absurd, but it is at least impractical and undesirable. And as far as the transit of passengers and mail is concerned, they can be conveyed by railway more economically. If the Suez Canal is to be attempted at all, which I hope it will not, I trust it will not be with English money."

De Lesseps was furious with Stephenson and the Prime Minister. He felt that Palmerston had called him, in public, a confidence trickster. He couldn't demand an apology from such an important man, so he did the next best thing. He sent a formal letter to Robert Stephenson challenging him to a duel. He wrote in French, but so that Stephenson would not be in any doubt of the nature of his challenge, he included an English translation. Fighting a duel was not exactly in Robert Stephenson's style – he quickly wrote a note of apology to de Lesseps and everything was smoothed over.

But de Lesseps left England and set up his own International Finance Company for the canal, without British backing. In Egypt the Viceroy bravely put up enough money to get the work started. Perhaps if people saw that de Lesseps and the Government of Egypt were in earnest, then wealthy investors, maybe even from England, might be tempted to take the risk. So, in April 1859, on a barren and uninhabited stretch of the Mediterranean coast, in front of a small group of Egyptian

Workmen loading camels in 1869. All early work on the canal had to be done without mechanical aids

workmen, Ferdinand de Lesseps lifted a pickaxe and struck the first symbolic blow to begin his canal. The long years of arguing were over. "It's only a little ditch now," he said with pride, "but it is the beginning of the Suez Canal."

He knew exactly what difficulties he had set himself – after all, he'd had many years to study them. The route he planned lay through a wilderness of shifting sands and desolate salt marshes, a string of vast dried-up lakes and a waterless desert. From Port Said, which he had yet to build, through a hundred miles of sand to Suez, a mere village in the burning sun on the coast of the Red Sea. De Lesseps knew that his most important task was to ensure supplies of fresh water and food for his workers. Stephenson's warning that the scheme would fail in a waterless desert echoed in his mind, for within six months of work starting, Robert Stephenson, a man of his own age, was dead.

At Port Said, Ferdinand built a stone jetty, a distillation plant for fresh water and a lighthouse. At the same time, he built a small canal to bring fresh water from the Nile close to the main channel. All the machinery had to be dragged across the desert by mules and camels, until the "ditch" was wide enough for de Lesseps to install the first of his sixty dredgers.

Wherever a ridge of rocks lay across the path of the canal, dynamite was used to blast a way through, for de Lesseps had planned that his ship-canal would have no locks to delay the big ocean-going vessels he hoped would use it. After three years of hard work, a narrow channel of sea-water had reached as far as the first of the dried-up lakes – Lake Timsah. De Lesseps decided that the time had come to stop work and hold a small ceremony. Saïd, the Viceroy, was invited to attend. Then de Lesseps stood close to the rough wooden barrier which barred the opening of the channel, and like Moses he called out: "In the name of the Viceroy and by the Grace of God, I command the waters of the sea to flow into the lake."

The barrier was moved, and the waters of the Mediterranean came tumbling into Lake Timsah. It was a fine sight – but it was the last time de Lesseps' friend Saïd was to enjoy a visit to the canal: he died soon afterwards.

His successor, Ishmael Pasha, took no interest in the canal scheme at first. Money and labourers were withdrawn, the work slowed to a halt. De Lesseps somehow managed to persuade the Viceroy to change his mind, and before long Ishmael Pasha began to call the Suez Canal *his* canal and became more eager for its success than anyone else had ever been. Wisely, de Lesseps named the most beautiful town on its banks after him – Ishmaelia.

At Ishmaelia the fresh-water channel from the Nile reached the ship canal, and wherever fresh water arrived, so did trees and fields and gardens. So at Ishmaelia the Viceroy built himself a summer palace where he could show off *his* canal to important visitors. And de Lesseps built himself a small villa there, too, so that his two sons – who had been left behind in Paris – could

come and live with him. And there he planted roses in his garden.

When the canal was only half completed, de Lesseps took the British Ambassador on a personal tour. He was duly impressed by what he saw: "It is no longer merely a de Lesseps' dream," he wrote. "The work is being pursued with striking energy and enthusiasm. We should not underrate what he has done."

Work was going on with increasing enthusiasm at every point along the canal. The dredgers were moving two million cubic metres of soil a month. The channel was deepened and widened, the banks shored up with stones. When the fresh-water channel arrived at Suez, the desert bloomed. From a village of five thousand people it blossomed and flourished to become a modern town of twenty-five thousand inhabitants.

Port Said grew as well, with hotels and shops for visitors; repair docks for steamers, and even a hospital for the canal workers.

In 1869, ten years after the work started, de Lesseps and Ishmael were able to announce to the world that the Suez Canal was complete. The Empress Eugénie of France had agreed to

De Lesseps' villa – now preserved as a museum – is seen on the left of this engraving of Ishmaelia in 1869

open it officially. She would sail down the whole length of the canal from Port Said to Suez in the Imperial Yacht.

Ishmael, the Viceroy, was tremendously excited by what had been achieved. He sent invitations out to a thousand people and all the crowned heads of Europe to come to Egypt and attend the opening ceremony at his expense.

In Cairo, he built roads, hotels, and he redecorated the city. He built a new Opera House and commissioned an opera. The official

The canal near Kantara in 1869, showing a dredger at work

opening of the Suez Canal was all set to be a very grand and very expensive occasion.

But not quite everything went according to plan. The large store of fireworks ordered for the opening ceremony at Port Said, suddenly exploded, long before the display, and two thousand soldiers rushed in to help put out the fire.

During a final check to make sure the canal was in perfect shape before the big day, a rock was found half-blocking the channel about twenty miles north of Suez. Men were sent to Cairo at top speed to buy gunpowder. "Buy as much of it as you can," de Lesseps told them. "If we can't manage to blow the rock up in time, we will at least be able to blow ourselves up." The channel was successfully cleared.

When the opening day came, everything seemed to be perfectly prepared. The Empress of France, the Emperor of Austria, and Crown Princes and Princesses from all over Europe arrived at Port Said and walked in procession towards two specially-prepared altars. The waters of the canal were to be officially blessed, twice. Side by side, Christian and Moslem ceremonies were solemnly performed. That night, Ishmael gave a formal banquet for his guests and there were fireworks – a fresh supply.

At eight o'clock the next morning, the flotilla of ships carrying the important visitors was due to start in procession down the canal. But at midnight there was a fresh crisis. Fourteen miles downstream, a brand-new Egyptian naval frigate had accidentally got stuck across the channel, in the dark, blocking all traffic.

Ishmael and de Lesseps were aghast. What could they do? They couldn't put off the Royal journey. Port Said was alive with rumours, and a decision had to be made quickly. If the frigate couldn't be moved, it must be blown up and all traces of it cleared away by the morning. Another naval ship left Port Said that night with top-secret orders and a supply of gunpowder.

Next day, not knowing what to expect, Ferdinand and Ishmael Pasha stood beside Empress Eugénie as her ship slowly steamed into the canal. And behind, in a long line, came fifty ships from every country in Europe. What would they find fourteen miles ahead? Would the channel have been cleared in time? Ishmael and Ferdinand de Lesseps smiled with relief when they saw that the frigate had somehow got clear. Their guests murmured in admiration at the sight of her, drawn up at the side of the Canal and ready to fire a salute of welcome.

The sun was setting when they reached Ishmaelia. The crowds were out in their thousands to cheer the ships and their passengers, who were to spend the night in the little town. Eugénie and Ferdinand were side by side watching, and Eugénie was crying with pride and happiness. A thousand Bedouin families in a thousand tents prepared to give the visiting Europeans an Arabic celebration they would never forget.

Next morning, Eugénie visited de Lesseps in his little villa and picked roses in his garden. She gave him a memento of the occasion – a little silver cup with her name on it. Then, leaving half the visitors still enjoying themselves in Ishmaelia, Eugénie continued her royal journey to the end of the canal. When they reached Suez, she officially presented Ferdinand de Lesseps with the Grand Cross of the Legion of Honour.

It was his moment of complete happiness. At last, the Suez Canal was on the map of the world.

He chose this moment to get married again. He was sixty-five years old, and Hélène, his young wife, was a girl of twenty; but they lived together very happily to the end of his days.

He came to England, where now everyone sang his praises,

Procession of ships down the canal

English celebrations for de Lesseps when he visited London in 1870 – a grand fête and firework display was held at the Crystal Palace

even Her Majesty's Government. Queen Victoria herself awarded him the Star of India.

Ferdinand de Lesseps was the hero of the hour. He had turned the dream of a lifetime into reality. His young wife gave him a happy home and they had a large family. Perhaps he should have slowed down and just enjoyed his success; but he didn't.

If he could join two seas together at Suez, what about Panama? Why not build another canal across that narrow strip of land which links North and South America, and that way join together two oceans – the Atlantic and the Pacific? He was about to start all over again, on plans for yet another canal. His son Charles, who had worked with him at Suez, tried to stop him. "What are you going to look for in Panama?" he asked his father. "By a miracle you succeeded at Suez. In one life isn't it enough to have one miracle without hoping for a second?"

But his father had decided. His reply to his son's pleadings was typical. "If you ask a General who has won his first victory whether he wishes to win a second, would you expect him to refuse?"

The French Emperor Napoleon III had been deposed and he and Eugénie spent the rest of their days as refugees in England. At Suez, de Lesseps had once called Eugénie the "guardian angel" of his canal; he had no guardian angel in 1879 when he and Charles took on the challenge of Panama. It was ten years since his triumph at Suez. But at Panama the problems were much more terrible than any he had met in the deserts of Egypt. Mountains, tropical rain-forests, and a sea of mud wherever the workmen dug. And disease, a deadly tropical disease that nobody could cure.

They struggled on for years. But de Lesseps was getting old, and he couldn't understand why year after year his best engineers all became ill and died. He thought it was cholera, and he had fought battles with cholera before. He built a modern hospital, but even the nurses died. It wasn't cholera, it was yellow fever – a swift and fatal disease. Twenty thousand people died, and in 1889 the Panama Canal Company went bankrupt. The jungle soon grew over his work, hiding all traces of it. The building of the

An oil tanker passes through the canal in the twentieth century

Panama Canal had to wait till medical science learnt to conquer yellow fever. The Americans eventually constructed it, in 1915.

Ferdinand de Lesseps, the general who had been so eager for a second victory, had failed dismally. He was accused of fraud; a warrant was issued for his arrest. Ferdinand was ill, too ill to attend his trial, but his son Charles was sentenced to five years' imprisonment. Ferdinand was nearly ninety, a pathetic shambling old man, and he retired to the country. He begged to be allowed to see his son Charles once more. Two prison officers stood outside the door when Charles de Lesseps came to visit his dying father.

The great de Lesseps, once such a hero, died penniless and in disgrace. All he had to leave his children was the little silver cup with the name Eugénie inscribed on it. It was just twenty-five years since his triumph at Suez. But there at least he was never in

The statue of de Lesseps which once dominated Port Said

disgrace. At Port Said they built an enormous statue of him that towered over his canal. Thousands of ships were using it, and the Suez Canal Company was far from going bankrupt. The British Government bought shares in it and were delighted. Over the years, the Canal Company shareholders made millions of pounds in profit.

Fifty years later, the Egyptian people had their own revolution; they deposed their king and decided to take over the Suez Canal and its profits themselves. The French and British tried to stop them, but the crowds swarmed into Port Said, took over the Canal offices and de Lesseps' great statue was sent toppling down into the water.

But that was more than twenty years ago, and the Suez Canal itself remains. Now, it's been made wider and deeper, so that giant oil-tankers from the Gulf States can travel through it. It's just as de Lesseps always planned it should be – every day ships from all over the world travel through his canal from sea to sea.

Looking back on the lives of de Lesseps, the Brunels, the Stephensons and Thomas Telford, it is their energy and their enthusiasm that seem so extraordinary. They were prepared to work so hard, to argue with anyone, to persuade Prime Ministers and fight governments, just to achieve the ideas they believed in.

They had to stand up to the people who condemned their dreams as impossible, and go on to break new ground. They were great men, who really cared about what they were doing.

When Thomas Telford built new roads so that people could travel safely between the towns and villages of Britain, he claimed that freedom demanded the freedom of good communications. When Ferdinand de Lesseps was pleading for approval to build the Suez Canal, he said: "Let us open the world to the people and break down the barriers that still divide men, races and nations."

That was the watchword of them all – these men of vision who helped build the world that we know today. They made the breakthrough, and we can still marvel at their skill, their daring – and their hard work.